'A fluid, engaging narrative.' *BBC Focus*

'It is rare to find an accessible book on the subject by a researcher who is an acknowledged leader in the field. Kent Kiehl is the exception to the rule... infectious... an insightful and memorable look into the troubled minds of people with psychopathy.'
Times Higher Education

'A lucid and closely observed portrait of what psychopaths are actually like, with their chilling combination of moral apathy and charm, by one of the leading researchers and innovators in the field. A fascinating and terrifying book, and a potential life saver.'
John Seabrook, The New Yorker

'A world-renowned researcher of psychopaths delves into the origins of their behaviour, especially as it relates to the inner workings of the brain... [Kiehl] navigates these issues and more with compassion and insight. Fast-paced and thrilling.' *Kirkus Reviews*

'A renowned neuroscientist provides us with a fascinating account of his personal journey into the mind of the psychopath.'
Robert D. Hare, author of Without Conscience

'A must read! Dr Kiehl's stories about psychopaths are as authentic as can be. His research formed the backbone of my work in analyzing the behavioural signature of a psychopath left behind at the crime scene. Whether you work in mental health, law enforcement, the judiciary or research, don't trust your gut about this disorder – learn about it from one of the foremost experts in the field.'
Mary Ellen O'Toole, Special Agent (retired),
Analysis Unit, FBI

Dr Kent Kiehl is Executive Science Officer of the nonprofit Mind Research Network and Professor of Psychology, Neurosciences, and Law at the University of New Mexico. He is frequently invited to deliver lectures to judges, lawyers, prison officials and lay audiences. The author of more than 100 scientific articles published in peer-reviewed journals, he has also been featured in media ranging from *The New Yorker* to *Nature*. He lives in Albuquerque.

The Psychopath Whisperer

Inside the Minds of Those Without a

Conscience

Kent Kiehl

ONEWORLD

A Oneworld Book

First published in Great Britain and Australia by
Oneworld Publications 2014
This paperback edition published in 2015

Published in the United States by Crown Publishers, an imprint of the
Crown Publishing Group, a division of Random House LLC,
a Penguin Random House Company

A CIP record for this title is available from the British Library

ISBN 978-1-78074-689-0
eBook ISBN 978-1-78074-540-4

Book design by Lauren Dong
Illustrations by Fred Haynes
Cover design by shepherdstudio.co.uk
Printed and bound in Great Britain by Clays Ltd, St Ives plc

Oneworld Publications
10 Bloomsbury Street, London WC1B 3SR

For mom and dad

Contents

The Psychopath Whisperer

chapter 1

Maximum Security

FACT: one in four maximum-security inmates is a psychopath.

Day 1

The snap of the lock releasing shattered the still morning air as the large metal gate, adorned with rows of razor wire, crept open along an iron rail. The 'lock shot', as it was known, echoed off nearby buildings, amplifying the eeriness of an already macabre scene. Two twenty-foot-high parallel chain-link fences stretched a quarter mile in either direction of the gate. In the space between the fences was an eight-foot-high column of razor wire, a gauntlet not even the most agile convict could vault. There was not a soul in sight. The gate appeared to mysteriously recognize someone was approaching and opened to welcome me to my first day working in a maximum-security prison.

That morning I had driven sixty miles through the rain from my residence in Vancouver to the town of Abbotsford, the home of several high-security prisons in the lower mainland of British Columbia, Canada. The Matsqui complex is just minutes off the freeway, surrounded by a collection of petrol stations and delis, no doubt to feed the hundreds of vehicles and staff who commute to the location each day. The entrance to the compound is nondescript except for the sign indicating that all visitors and vehicles on the property are subject to search and seizure for contraband. The vista stretches as far as the eye can see, rolling hills of dark green grass dotted with castlelike structures surrounded by moats of high fences topped with razor wire and fifty-foot-high turrets placed strategically at each bend in the fence. At the end of a long road is the Regional Health Centre (RHC) – a name that belies its guests. RHC is a maximum-security treatment facility for sex offenders and violent offenders. Its 250 beds contain some of the most dangerous criminals in Canada. It was my new place of work.

I was a twenty-three-year-old first-year graduate student. On the early morning drive, I thought about how wholly unprepared I was for my first day of interviewing prisoners in the violent offender facility. For the past several years, I had divided my time between studying the research literature on psychopaths, undergoing training in brain-imaging techniques, and engaging in a loosely related line of research on a study of the brain electrical activity associated with auditory processes of killer whales, comparable in many ways to those of humans. Becoming ever more fascinated by psychopathy research, I had also been vigorously pursuing mentorship with my academic hero, the founding father of modern research in psychopathy, Professor Robert D. Hare, who only recently had accepted me as a graduate student. Yet now, as I walked past the metal detectors at the entrance to the compound, surrounded by razor wire, I paused and wondered what the hell I was thinking. I would be working, all alone, on the forbidding task of conducting in-depth interviews with the prison's most violent inmates, many of whom had been assessed as psychopaths. After the interviews, I planned to administer EEG (electroencephalogram) tests, measuring electric impulses in the brain in response to emotionally loaded words – data that would help us understand the connections between psychopathic brain processes and behaviour.

I cleared security, received my ID card, and was given directions to the department of psychiatry by a guard, pale and gaunt, who looked like he had spent fifty years behind bars. The next lock snapped open with a now familiar audible crack and the heavy lead-lined door popped open; I gently pushed it forward. As I took my first few steps into this new environment, I smiled to myself that my first concern, a cavity search, had not come to pass as I went through security. I made a mental note to get even with the senior graduate student who had told me that cavity searches of new staff were common in Canadian prisons.

Inmates dressed in white T-shirts, jeans, and dark green jackets milled around the laundry, barbershop, and chapel as I walked from the administration entrance to psychiatry. The halls smelled like disinfectant, and I pondered what chemicals were used to clean up blood.

I entered the large, ominous building at the end of the walkway. I wandered down the corridor like a lost child until I came upon a sign on an office door that said DR BRINK. Sitting there, oddly facing away from the open door, almost inviting me to sneak up and scare him, was Chief Forensic Psychiatrist Dr Johann Brink. I'd met Dr Brink just three months earlier at a NATO-funded Advanced Study Institute on psychopathy in Alvor, Portugal. Over numerous dinners and bottles of wine, I had convinced Dr Brink to collaborate with me on my EEG studies of psychopaths. He helped me get my protocols approved by the prisons and the university ethics boards. With all this paperwork in hand, I tapped lightly on the door frame to his office. He spun around, only partially startled, and greeted me with a huge smile.

'Kent, great to see you! Welcome to maximum security!' he bellowed in his distinctive South African accent.

Johann proceeded to walk me down the corridor and show me my office, empty except for a phone and a desk with two chairs on opposing sides. A large, bright red button was positioned chest high, right in the middle of the wall.

'I recommend you take the chair closest to the door; just in case you piss one of them off, you can run out quickly. Better than getting caught on the other side of the desk. If you can't get out, hit that red button and the guards should come running.' He spoke so casually I could not help but wonder if he was kidding.

'And here is your key – don't lose it!' I was handed a six-inch brass key with large, odd-shaped, forbidding teeth. The key, made by only two companies in the world, was specific to prisons. It opened most doors.

He pointed down the corridor to a large door. 'The guys' cells are through there. I've got to run now; we can check in at the end of the day, eh?' Johann smiled and turned away as he was finishing his sentence. As I inserted my new key into the shoulder-height lock in the door, I faintly heard him say what I thought was 'Enjoy!' as he closed his office door tightly – no doubt to keep the next visitor from sneaking up on him.

I pushed open the hallway door to the prisoners' cells, turned and closed it, inserted my key on the other side, and spun the heavy-duty lock 180 degrees. I tugged on the door to make sure it was locked, took a deep breath, and proceeded down the 100-foot-long hallway to the inmate housing units.

I arrived at a 'bubble' – a round security room, with one-way, tinted windows, no doors apparent. I gazed down the four corridors that radiated from it like spokes, as maximum-security inmates milled about, staring coldly. I wasn't afraid, but rather nervous about whether anyone would agree to talk with me. By way of training, Professor Hare had handed me a worn copy of the introductory book on life in prison titled *Games Criminals Play* the previous day and said, 'Read this first, and good luck tomorrow!' It was trial by fire, sink or swim. *I should have read the book last night,* I thought.

A small office with glass windows and a half door was across the hall; a forensic nurse was handing out shaving razors to a queue of inmates. She looked at me curiously and waved me over.

'Can I help you?' she asked cautiously.

'I'm the new research guy from UBC. I'm here to sign inmates up for interviews and EEG studies.' UBC stands for University of British Columbia, where I was a doctoral student studying psychology and brain science. EEG stands for electroencephalography (also electroencephalogram), or the recording of brain electrical activity using noninvasive sensors attached to the head, amplified, and recorded on a computer for subsequent digital analyses.

'Come in and sit down then; let's talk about it.'

I leaned over the half door and looked for a latch.

'On your left,' she said. I found the latch and flicked it open and sat down in the closest chair. She finished handing out shaving razors to the last of the inmates and turned to look at me.

'The inmates get razors?' I inquired with a puzzled tone.

'Yes,' she said, laughing, 'and they often disappear; I don't ask where they go.'

I realized that Dr Brink was wise to tell me to sit closest to the door during my interviews.

Dorothy Smith was a twenty-year veteran of the maximum-security prison. Despite her long stay in prison, Dorothy was no worse for wear. Her slim athletic build was topped with an infectious personality that won over even the most hardened inmates. She would become one of my closest friends during my seven-year term in Canadian prisons. And she shared my interest in figuring out what made psychopaths tick.

'I'll set you up with a nice one for your first interview,' Dorothy said as she glanced up to the housing chart taped to the cabinet. I followed her gaze and noticed headshots of inmates on the cabinets behind me, their names listed underneath: last name, first name, and index crime. Attempted murder, rape/murder, arson/murder, murder 3x, murder/rape. I pondered if 'rape/murder' and 'murder/rape' were the same thing and was about to ask Dorothy when I thought better of it. I didn't want to know; I had enough on my plate for my first day.

The inmate she selected, 'Gordon',* seemed courteous enough as he sat down in the chair on the far side of my office. He was forty-two years old, balding, grey haired, and soft spoken; the crime listed under his headshot was 'attempted murder'.

A fascinating guy, Gordon turned out to be a serial bank robber. He told me his crimes had financed a lavish lifestyle, including first-class international plane tickets, front-row seats at hockey games, and girlfriends and prostitutes in many different cities. After his most recent arrest, Gordon had to explain to the police why he had

---

* I have changed the names and characteristics of my interviewees to protect their identities.

more than $75,000 in cash, despite being technically unemployed. With his lawyer, he negotiated immunity throughout Canada on the condition that he help police clear up a number of unsolved bank robberies. The number of robberies that Gordon was directly involved in reached close to fifty, but he was never charged in any of them. Gordon regaled me with story upon story of successful bank robberies. He told me how to case a town or city, then the banks, how to get in and out in less than sixty seconds, how to steal a getaway car, and how to launder the money. I asked him how a bank could keep him from robbing it? He gave me hours of insights. I started making notes about how to design a better bank – *Perhaps,* I thought, *I can consult with bank executives if my academic career doesn't work out.*

My interview with Gordon was designed to cover all domains of his life. We reviewed his upbringing, education, family, friends, sporting activities, work experience, career goals, finances, health, intimate and romantic relationships, substance abuse and impulsive behaviours, emotions, antisocial behaviours, and his index offence. The interview typically takes anywhere from one to three hours. With Gordon I spent six hours. We pored over all the details of his life. If if I hadn't been hooked on this career path yet, I was after speaking with Gordon.

Our review of his work experience was brief. Gordon had had dozens of jobs, but he never held one for more than a month. He was routinely fired because rather than working, he preferred to play jokes, take long lunches, drink, and gamble. Most of his jobs were in construction or as a car mechanic (he admitted to choosing this vocation so he could become a better car thief). When asked about his future plans, Gordon said he wanted to leverage some of his residual bank robbery proceeds to start a motorcycle dealership. He failed to appreciate the potential legal (and tax) implications of such an endeavour.

During our discussion of his finances, Gordon admitted he rarely used bank accounts.

'Afraid of someone stealing it?' I quipped.

'No,' he replied with a wry smile, 'I just don't like to have to explain to people where I got the cash.'

'If you don't keep your money in a bank, where do you keep it?'

'I bury it,' he said with a laugh, 'all over the place; you can't just drive around with hundreds of thousands of dollars on you after a job. Or I FedEx it to a five-star hotel in Asia, Europe, or South America where I have a reservation under a false name; once in a while I will FedEx it to a girlfriend in another town, tell her it's a present but not to open it until I get there, stuff like that. You always have to be careful when you get to the hotel to make sure the cops are not on to you. I usually wear a disguise and scope out the place. I like to send the package two-day overnight and fly there first, then watch the delivery to make sure there are no cops. I've only lost a few packages.' He paused, then laughed again as he told me about a prostitute he had sent $50,000 in cash to a few years back. She never picked him up from the airport and he never saw her again. 'I knew better than to trust that girl.'

'Money belts used to work through airports too, but it's harder to do that now, more risky,' he mused. He continued. 'Mules sometimes, but really, I just like to go for a hike and bury it. Then I know it's always going to be there when I need it. I have lots of good places to bury it, but I don't talk about that stuff with anyone.'

We turned to talking about his views on relationships, family, and friends. Gordon was a loner; he'd never felt the need to be close to anyone. He'd had hundreds of sexual liaisons, starting at the age of eleven. When asked if he had ever been in love with someone, he replied quickly and with a large smile about the time he was with three prostitutes at the same time – for a week.

'Ah, I loved them all,' he said as he took a deep breath, remembering.

Gordon equated love with good sex. He'd been married six times, all in different countries, all under aliases. When asked why he got married, he replied: 'Makes the girls happy, keeps the sex coming for a while, and they are more willing to help mule or receive stolen money.' Gordon admitted the prostitute he sent the $50,000 to and never heard from again was wife number three.

He had not talked to his parents or siblings in years; the last time he bumped into his sister he heard that the family was fine, living the 'white-picket-fence dream'. Inmates commonly refer to the presumed tedium and boredom of having a simple job, wife, and house as the 'white picket fence' syndrome. Actually, the majority of inmates admit they would be very happy out of prison and residing behind the white picket fence; psychopaths, however, can't fathom it – they laugh at others who would dwell in such monotony.

Gordon viewed others as untrustworthy, but he was affable, engaging, quick witted, and full of stories. I didn't believe all the stories – lying is common in psychopaths – so I had to trust my gut feelings and review his file again later. Because you can't trust a psychopath is telling you the truth, you have to carefully review all their files in order to be able to verify everything they say. If you catch them in a lie, you have to be willing to call them on it and see how they respond. *Just sit in the chair closest to the exit – in case you piss them off.*

After another hour discussing his upbringing, I started asking questions about what Gordon liked to do when he was a child. He grew up in Abbotsford, not far from the prison where we now sat. Nestled up against the mountains in British Columbia, Abbotsford has wonderful views of Mount Baker to the southeast, and the local mountains are full of excellent fishing, hiking, and mountain bike trails. Gordon told me stories about his favourite fishing spots, remote lakes and places with great views. He started bragging about the enormous fish he caught – lies, I figured. He liked to hike alone even when he was a kid. After talking about a bunch of his favourite spots to go as a child, he abruptly stopped, looking at me and then the video camera recording our interview.

He said, 'That was good. You are nobody's fool. You got me to talk about all the places I liked to go hiking and fishing as a kid so you can try and figure out where I buried the dough. People have been trying to get me to talk about that for years. You're good.' He laughed.

I also laughed and told him that getting all the details of his childhood was a necessary part of the interview. I was bluffing. As soon as he told me he buried his bank loot, I had been tailoring the interview to find a way for him to give up the locations. I knew Gordon had spent much of his adult life in prison. I figured the locations he selected to hide his treasure had come from his childhood. Also

I leveraged his abundant grandiosity to get him to tell me stories of the 'big' fish he had caught.

My main motivation was to perfect my interviewing techniques – could I successfully get a subject to talk? But a part of me couldn't help but think about how I could train my black German shepherd to sniff out cocaine. A high percentage of $20 and $100 bills have trace residue of cocaine on them from the drug trade. It certainly would make hiking more exciting if your dog was able to dig up a bag full of cash!

I finished my interview with Gordon by asking him to describe his latest offence. It seemed that he was quite good at robbing banks, but not so good about keeping his temper in check. Gordon had 'poor behavioural controls', another classic trait of psychopathy; he was constantly getting into verbal altercations that escalated into physical fights – with little or no provocation. His current index offence was for almost killing a lover of one of his girlfriends. He had suspected she was seeing someone else, followed her, and confronted them. Tempers flared and Gordon wounded the other guy with a knife. He was arrested the next morning. It was noteworthy that, in prison, Gordon was pretty well behaved. He knew that fights in prison often led to hard time and additional charges, so he kept his temper in check. He wanted to get out as soon as he could so he could dig up his cached treasure.

When Gordon left, I pulled out my copy of the *Manual for the Hare Psychopathy Checklist-Revised* (PCL-R).[1] The Psychopathy Checklist, created by Professor Hare, is the instrument we use in the field to assess psychopathy. It contains twenty items that capture the essential traits of psychopathy – including lack of empathy, guilt and remorse, glibness, superficiality, parasitic orientation, flat affect, irresponsibility, and impulsivity. These traits are assessed based on the individual's entire life and in all domains of his or her life. That is, to 'lack empathy' on the Psychopathy Checklist, you must have evidence of this trait in the majority of your life – at home, work, school, with family, friends, and in romantic relationships. Each of the twenty items is scored on a three-point scale: 0, the item does not apply to the individual; 1, item applies in some respects; and 2, item definitely applies in most respects to the individual. The scores range from 0 to 40, with the clinical diagnosis of a psychopath reserved for those with a score of 30 or above. The average inmate will score 22. The average North American nonincarcerated male will score 4 out of 40.

Gordon scored 31. He was my first psychopath.

I finished my notes justifying my Psychopathy Checklist item scores on Gordon while I wolfed down the two peanut butter and jelly sandwiches I had packed the night before. I needed energy to continue with my next interview. It's taxing to focus for hours interviewing inmates, ever cautious about walking the line between getting the information you need, challenging them to be forthright, and monitoring the door to make sure you can make a speedy exit if need be.

I returned to the housing unit and was approached by Gordon's roommate, 'Grant', my second interview. Gordon had told him that I was fun to talk with and to 'try me on'.

Grant had the kind of conventional appearance and manners that would suit a car salesman, except for the bold, spiralling tattoos covering his arms and hands. He had been involved with the legal system since birth – his mother was incarcerated when he was born – and he was currently finishing out a fifteen-year sentence for two murders he committed at age thirty. Grant was charged with manslaughter in the killing of his two accomplices to a robbery. Apparently, a disagreement occurred about the splitting of the proceeds. Knives were flashed but they were no match for Grant's 9mm handgun.

'Bam, bam ... bam, bam ... two down,' he said, pointing his index finger and thumb in a classic gun pose. 'One of my better shooting days.' The killings were spoken of with such calmness, such 'matter-of-factness', that I wondered if they were true. The files confirmed it; two shots centre-mass on both accomplices. Grant received fifteen years to life for the slayings, largely because of a plea deal – there was not enough evidence to convict him of first- or second-degree murder and he had disposed of the weapon. Interestingly, he was distraught when he talked about getting rid of the gun, his favourite, a Glock 17 with extra magazine capacity. He had decided to plead out the case when his lawyer told him that the prostitute he had hired to be his alibi witness was likely to break down on the stand.

When I asked if he had done anything for which he hadn't been caught, Grant laughed with what seemed like childish mischief, and said, 'Lots . . . arson, robberies, breaking and entering, car theft, check and credit card fraud, and of course there are a few bodies around.' He'd shot a few strangers, he said, for getting in his face, and drowned at least one girlfriend in a pool . . . which is when I realized I was sitting across from a bona fide serial killer, albeit an extraordinarily friendly one.

I eventually got Grant to concede that he had ten murder victims. Oddly, he'd never counted them up; in fact, he hardly ever thought about them. I tried to place Grant's murders within the context of a classic serial-killer profile. It didn't fit. Most serial killers are driven to commit their murders, usually in association with sexual dominance or sadism. Serial killers like Ted Bundy would meet criteria for psychopathy *and* they also have a paraphilia (a sexual-based disorder), like sexual sadism. The drive to kill comes from the latter; the lack of emotion, empathy, and guilt comes from psychopathy. When you combine psychopathy with a paraphilia, you get a very dangerous person. Fortunately, such people are very rare.

Grant didn't have any sexual disorders like sadism. He described a relatively normal sex life. Yes, he admitted, he'd been a little rough from time to time with women, but he didn't get turned on by inflicting pain. In fact, most of Grant's murder victims were male. He seemed to resort to violence easily, quickly, and without much thought. He was lucky to have been caught for only two of the ten murders.

The rest of my interview with Grant revealed he'd been getting in trouble since his early teens, had been arrested many times, had been in lots of fights, and was unable or unwilling to stick to any occupation, profession, or job for more than a few months. He had collected social assistance under multiple names, been married a few times, and had four children – as best he could remember. This latter point was quite interesting. Grant didn't recall the birthdays of his children; in fact, he knew only two of their names. He'd led a nomadic existence, moving from place to place – often on no more than a whim – living out of his van, camping, occasionally shacking up with women, occasionally getting them pregnant, and always moving on to the next adventure.

Psychopaths rarely know details about their children. Like Grant, they often don't even know how many children they (might) have. I would come to realize during my research that psychopaths' lack of connection with their children is one of the most salient features of the condition.

It was getting late in the day, so I wrapped up my interview with Grant, indicating to him that I might want to do a bit of follow-up later. He stood up, reached forward with his hand extended, and we shook hands. 'Let's do this again,' he said. 'It was fun.' And he walked out as if he had just given a press conference.

A bit confused, I sat down and pulled out my PCL-R manual to assess Grant on the Psychopathy Checklist. I leaned back in the chair and, for a moment, wondered if I was dreaming. The day had been surreal; after all the years of reading about psychopaths, I'd finally conducted my first two interviews.

Grant scored 34 on the Psychopathy Checklist. I was two for two.

I finished up my notes and dumped the videotapes of the interviews into my locked file cabinet. We record all our interviews so that another scientist can rate the psychopathy scores too. In this way we 'double-rate' everything, making sure that the interviewer did not get a biased impression of the subject.

I closed and locked the door, returned my monster key to the nearest lockbox (the brass keys don't leave the facility), and walked down to Dr Brink's office.

Dr Brink's door was closed; he was busy typing up notes on his computer, still facing away from the door. I tapped softly on the window. He turned, smiled, and got up to let me in.

'How did it go today?'

'Fascinating. It's just like I imagined, only better,' I told him. 'No problems. All is going well. I got to meet Dorothy, did two interviews, and got leads on more inmates to interview tomorrow. I'm ready to head home for a beer.'

'Excellent. Well, don't let me keep you from that beer. Have at it the rest of the week and stop by and see me if you need anything,' said Brink.

I wandered back out the maze of corridors, past the chapel, laundry, and barbershop. All were empty at 6 p.m., the inmates already locked up for the night. It was quiet, almost peaceful – until I was startled by the lock shot of the final gate opening. I passed through the gate and was overcome with a feeling of freedom. I would never forget my first day in prison. Nor, I would discover in the years ahead, would I ever walk out of a prison and not feel a small sense of relief to be back on the outside.

I climbed into my Toyota pickup truck and began the long drive to Vancouver. That day, at the beginning of my career in the prisons, my truck had 40,000 miles on it. By the time I retired it years later, it had more than 280,000 miles on it – a *moon unit* as I called it, referring to the fact I had accumulated enough miles driving back and forth to prisons that I had travelled the equivalent of the distance between the earth and the moon (238,857 miles).

On the ride home, I couldn't stop thinking of how I might have given better interviews. I thought back on my interview techniques, trying to think of ways I could have got more details out of the inmates, how I could make the scoring of the Psychopathy Checklist easier. I realized that I needed to edit the default semistructured interview that comes with the Psychopathy Checklist, and I needed to find the right balance between asking more questions – probing to get all the details from different areas of an inmate's life – without making the inmates angry. By the time I pulled into my driveway an hour and a half later, I had come up with a dozen new questions to add to the interview.

My 110-pound black German shepherd, Jake, raced around the corner from the back garden to greet me as I opened the gate. Seeing him, I was reminded of my idea to train him to search for cocaine so we could track down the loot from my new bank-robber friend.

My housemate, Andreas, a starving artist and conservationist, returned home from his double shift at Starbucks.

'Did you bring the drugs?' I asked Andreas.

'Yup,' he said, pointing to the overstuffed rucksack he had dropped in the corner of the living room.

Andreas had been promoted a few months back to shift manager at Starbucks, and one of his responsibilities was to monitor the age of the coffee beans being used and sold in the store. He was told to throw away any coffee beans that had expired. But the beans were still quite good. He couldn't bring himself to throw them away, so he brought them home where they decorated the living room in our small apartment like beanbag chairs. Before long we had over a hundred pounds of coffee in our living room. I finally convinced him to give them away, and he started shipping them to family and friends all over Canada.

'How was your first day?' Andreas asked, a little nervously.

'Amazing. Interviewed my first serial killer.'

'I don't know how you do it – it would freak me out,' he said.

It's a common question – 'How do I do what I do?' Or even more often, 'What are psychopaths like?' But what questioners are really wondering is, 'How did I develop such an interest in psychopaths?'

I grew up in Tacoma, Washington, a couple blocks away from the house serial killer Ted Bundy was raised in. My father was a writer and lead editor at the *Tacoma News Tribune*, the local newspaper. I was just a kid when the story of Ted Bundy broke in the '70s; my father would come home and tell stories he'd just edited for the paper of the child from down the street. My family would sit around the dinner table and wonder how someone like that could grow up in our sleepy little middle-class suburb. How indeed? That seed just sat there in my brain, waiting to germinate.

I was not much of an academic in high school. I skated along with a B+ average, opting to put my energy into sports: American football, lifting weights, and track. My dad was a big reason I was successful in sports. My father never missed a single sporting event of mine – twelve years of baseball, ten years of football, four years of American football, and four years of track. A sportswriter, he could rattle off every stat from every professional baseball player. He was

born with a muscular condition l
lacked the fast-twitch muscle fibr
but that never left him short of ent
friends and me for years in baseb
school American football season, h
giving me insights into how to pre
face. He was as dedicated as my sc

My parents worked very hard
myself and my three sisters, into t
state. Bellarmine Preparatory High School was a place with highly
dedicated teachers and an amazing community environment.
Ninety-eight percent of my graduating class went on to a four-year
university, so it was peer pressure that got me to apply to university.
But it was sports that got me accepted.

I applied to a number of different universities and was recruited
to play American football at the University of Washington, Washington State University, and a few others. Don James, the University
of Washington football coach, told me that I could come play for
him, but that I would not likely start on a first-division university
team for all four years. James thought that with continued progress,
I would be a good wide receiver or free safety. I had started high
school a scrawny five feet, nine inches, 150 pounds, but I played my
senior year of high school football at six three, 205 pounds. James
told me if I wanted to be on the starting line-up all four years of
college, I might consider the top second-division programme in the
country – which at the time was the University of California, Davis.

So I sent my films to UC Davis football coaches and they recruited me. They also helped get my application accepted, and I decided to head off to California.

My athletic career was short-lived; my knee folded over at the
end of my first year at UC Davis. After rehabbing for a year, my
dream of catching footballs for a living officially died. I was strug-
gling to find something to sink my teeth into. I turned to Dr Debra
Long, professor of psychology and my undergraduate adviser. One of
the wonderful things about the academic environment at UC Davis
was that undergraduates were encouraged to work closely with professors. I had been able to work a great deal with Dr Long over my

Davis and she knew me very well. When I visited
of my second year seeking advice, she told me, 'Kent,
scientific mind; I want you to go away this weekend and
ck Monday and tell me five things you would love to study
ur life. I think you should consider being an academic.'

She told me I had way too much energy to pursue a nine-to-five
job. I needed a career, she said, not a job. I ruminated over the weekend and returned to give her my list of subjects I wanted to study: (1)
the brain, (2) psychopaths (inspired by my childhood curiosity about
serial killer Ted Bundy), (3) killer whales (another seed planted in
childhood when a killer whale looked me right in the eye while on
a fishing trip with my dad in the Puget Sound), (4) teaching, and (5)
women. She got a good laugh out of the last one.

Dr Long called a couple of other professors, Dr Michael Gazzaniga, the founding father of the field of cognitive neuroscience (the
study of how the brain processes information), and Dr George Mangun, an attention researcher, both of whom had just relocated their
laboratories from Dartmouth University to UC Davis. She told them
that she had a motivated undergrad she was sending their way. Next
she called Dr Carolyn Aldwin in the human development department. Carolyn was married to Dr Michael (Rick) Levenson, a research professor who studied psychopathy, among other conditions.
She also set me up to see a lecture by Michael Szymanski, a graduate
student who was studying brain electrical activity in killer whales.
Wow! Dr Long still receives free drinks anytime our paths cross.
All of the individuals she contacted that day became mentors and
lifelong friends, and eventually, I am honoured to say, I came to be
called a colleague by them.

My life was transformed. I had found my path forward in life.
I wanted to be a professor and learn everything I could about psychopaths. I wanted to master brain-imaging techniques and teach
the world what is different about these individuals, what's going on
inside psychopaths' minds.

I quit partying and became serious about my studies. I went from
a hundred buddies to three or four good friends. I went from a B
average to straight A's. I was advised that if I wanted to study psychopathy, I should do my graduate work with the most prominent

scholar in the field, Professor Robert Hare of the University of British Columbia.

And that's how I started down the career path that brought me to maximum-security prison.

Day 2

In the distance I could see the sun rising behind Mount Baker as I pulled into the car park of the Regional Health Centre. I grabbed my rucksack off the passenger seat and walked up to the gates.

The same ancient guard who had let me through security the day before waved me past the metal detectors without a second glance. I stopped and knocked on the bubble's window.

'You forget the way?' the voice heckled over the speaker.

'Nope. I thought you might like some coffee.' I reached up, showing him a one-pound bag of Starbucks breakfast blend (courtesy of my housemate, Andreas). The lock shot of his door fired instantly and he pushed open the door.

'Absolutely,' he said. Smile lines creased across his face. 'Thanks much!'

I pushed the heavy entrance door open and headed towards my office. I was not more than twenty feet into the prison when Grant emerged from the laundry carrying his bag of clean clothes.

'Hey, Kent,' he said. 'You got a second?'

'Sure,' I said, 'what's up?'

A worried expression appeared on Grant's face. After looking up and down the corridor and seeing no inmates, he said, 'I'm not sure what you did, but rumour has it that one of the sex offenders doesn't like you. His name is Gary. Just keep away from him. Okay?' He pulled away and went back into the laundry.

'Sure,' I managed to choke out.

My mind started racing with what happened yesterday. I couldn't think of anything I had done to piss off an inmate. It's the last thing I was trying to do.

The RHC is divided into a number of different housing divisions. On the west side of the complex is a wing of thirty or so beds for inmates who have a severe psychiatric illness, like schizophrenia. These latter inmates typically don't interact with the main population because they are very ill. The other housing units are contained within a two-level complex with four tiers radiating out from a central hub. The four arms of the first level house mentally challenged inmates. This includes inmates with low IQs or other mental problems. The top four tiers are split into two arms for the violent offender treatment programme, and two for the sex-offender treatment programme. Each of the arms houses twenty to forty inmates, depending upon whether they are double bunked or not. Each of the top-tier treatment programmes operates on a rotating schedule such that twenty-five new inmates turn over every three months as the nine-month treatment programmes conclude. This schedule provided a steady stream of inmates for my research studies.

Normally, sex offenders and other offenders are segregated from one another. This is done because in the prisoner hierarchy inmates who have committed crimes against women are scorned by the other inmates. Inmates who have sexually assaulted a child are considered to be the 'lowest of the low', and they are often victimized in prison. So for their safety and for the safety of the staff (who might have to try to break up assaults), sex offenders and other offenders are kept separate.

But at RHC, sex offenders and violent offenders share the same treatment schedule and are allowed to intermix. This sometimes leads to conflict, but because the offenders all volunteer to be part of the treatment programme, they are generally much better behaved than they would be otherwise. For most offenders, the RHC treatment programme is a stepping-stone to early parole, so their tolerance levels are fairly high for changes to the prison routine.

So during my time at RHC, I found it was not uncommon for 'vanilla' inmates to socialize with sex offenders.

I wandered up to my office in a daze. I could not understand what I might have done to set off an inmate.

I figured I would just try to avoid this Gary until I could figure out what to do.

I updated my interview schedule for the Psychopathy Checklist with a couple dozen new questions and printed off two copies. I grabbed my consent forms, a bag of coffee, and removed my brass key from the lockbox. I instinctively palmed the brass key, mimicking the self-defence technique taught to women to fend off attackers using car keys.

I headed down to the housing units and went straight into the nurses' station, not waiting for an invitation to enter. I just unlocked the door and walked in, making sure that no inmates came up behind me. I handed a bag of coffee to Dorothy. A smile widened across her face as she took the coffee and started a fresh pot.

'This is a nice surprise,' she said. Then she turned and looked at me. 'Something the matter?' Dorothy had world-class clinical skills honed after twenty years of reading the faces of inmates. I thought she must be a fantastic poker player.

'Umm, do you know a sex offender named Gary?'

'Sure,' she said. 'He's over there. He's a troublemaker, that one.'

I was not sure how to deal with Grant's earlier statement. I had promised him confidentiality, and I was not sure if the details he gave me could be shared with someone else. I decided to keep my conversation with Grant confidential, but I felt it safe to get details about Gary from Dorothy.

Dorothy volunteered that Gary was always into something. He caused a lot of problems with the team leader running the group therapy sessions; he had punched a hole in the wall recently after a shouting match with the therapist (apparently the treatment team was pleased with this outcome because it was progress for Gary, since he hit a wall and not the therapist); he'd been caught making brew (a concoction of yeast, fruit, and water that ferments into an alcoholic beverage and is usually stored in the inmates' toilets); and he had been suspected of beating up a couple of other inmates, although nobody had come forward with complaints.

I saw under Gary's photo on Dorothy's housing chart that he was serving time for two rapes with a sentence of fifteen years. He'd served fourteen years and was up for release soon. He was at RHC because the Canada Department of Corrections was trying to get him some treatment before they were forced to release him on completion of his sentence. I later learned that Gary had been rated as very high risk to reoffend, and the administration was trying to find a way to mitigate that risk.

Gary was gigantic. Many inmates are ripped from working out several hours a day, but Gary was well over six feet tall, barrel chested, with huge shoulders, and he weighed in at nearly 275 pounds.

I twirled the brass key in my hand . . . my anxiety increased a bit as I realized the futility of trying to stop 275 pounds of angry sex offender with a tiny brass key.

As Gary walked by on his way to group therapy, I could feel his eyes on me. I relaxed a bit once I saw him pass through the doors on his way to the therapy rooms.

Gordon meandered by, spotted me, and headed over to the station.

'I got some new guys signed up for you,' he said, reaching into his pocket and pulling out a crumpled piece of paper with over fifteen names written on it.

'Thanks for helping out,' I said.

'No problem. Hey,' he said, looking around and making sure no one could overhear. 'Do you think I could get a commission or something for helping you out?'

I smiled.

'Nice try. I'm afraid that the best I can do is let you go first on the next phase of the research. We are not allowed to involve the participants in the recruitment of other subjects. But I appreciate your helping out.'

Gordon replied, 'Oh, well. It was worth a try . . . so when's the next phase start?'

'A couple weeks. Just putting the equipment together now,' I said.

'Cool.' Gordon turned and walked away.

I looked down the list of names and saw that it appeared Gordon had gone cell to cell the preceding night and got his entire tier to sign up. Or more likely, he just wrote down the names of everyone in his cell block and never even bothered to talk to them. Had I become cynical after just a single day?

I exited the nurses' station, looking left to make sure the door through which Gary exited had not opened to let anyone back into the housing units, and walked down to the small common room at the end of Gordon's tier. Inmates were milling around. I called out the first name on my new list, 'Mike West. Is Mike around?'

'Yup. That's me,' a voice called out from the back of the room.

'You got a minute to talk about doing a UBC research study?' I asked.

'Sure,' came the answer. A tall, thin man walked over to me and said, 'Let's go. Gordon told us all about it last night.'

I began to think that I might have been too quick to judge Gordon's intentions or, rather, his motivation to get a commission.

'Mike' had a drug problem. He started out our interview by quickly volunteering that he wasn't as violent as the rest of the guys in here. He just did a robbery that got a little out of control, and people got hurt.

'I was just trying to get money for my next fix,' he recalled.

Mike grew up just outside of Toronto. He made it through high school and started working construction, gradually moving up the ranks until he was an operator of heavy machinery. He'd had a number of jobs, with the longest lasting over four years. He was pushing forty when I met him, having completed five years of a six-year sentence for aggravated assault and battery, evading police, reckless endangerment, and a few other charges. Mike was married going on ten years. His wife had moved so she could visit him frequently. She lived only a few miles from the prison and worked as a waitress in a local restaurant.

He started using marijuana, a 'gateway drug' he called it, in high school. Dealers started him off using, then selling, then asked him to try some other stuff. He tried heroin and was hooked right away. Mike tried to use recreationally, but eventually he was using regularly. He kept it hidden from family, friends, and even his eventual wife, until he had no way to explain where all his money was going. He started doing burglaries, pawning stolen TVs and electronics equipment. He fell into debt, took out loans, sometimes from the wrong people, and then started doing armed robberies. Simple stuff, he explained – rob a 'mom-and-pop' place and get four hundred to five hundred bucks, enough to buy a week's worth of drugs. He had tried to stop using, he told me, but it just wasn't in the cards. Then 'D-day', as he put it. As he completed a robbery of a mini market, he ran outside and jumped into his car. A cop car happened to be passing by and got the call. A high-speed chase ensued and ended with a crash. He put the family of the car he hit into the hospital with serious, life-threatening injuries. As he told the story towards the end of our interview, tears welled up in his eyes.

'I really messed up,' he said. 'But now I've been clean and sober for more than five years.'

Mike spent a year in jail prior to trial and being sentenced to the federal prison. In Canada, sentences less than two years' duration are served in provincial facilities, similar to a county or city jail. Sentences longer than two years deserve federal time and are served under the Canadian Department of Corrections, a nationwide entity with prison facilities in all the provinces.

Mike detoxed in the provincial jail and received psychological and pharmacological treatment – substance abuse cognitive behavioural therapy and medication to help with the withdrawal effects, the dosage of which was eventually titrated down until he was clean. He wasn't going back to that life, he had promised his wife and family. He was doing his time and going to go back and get a construction job and start a family. The 'white picket fence'. He didn't have to say it; it was clear that his dreams only went that far.

Mike was emotionally connected to his family, his wife, and even other inmates. He taught rudimentary maths and English to other prisoners and received good behaviour credits. He did not have a single institutional infraction, instance of drug use, or a fight. Mike steered clear of problematic inmates. He had been originally sentenced to a medium-security facility, but within a year had been transferred to minimum security. He was deemed to pose little escape risk and was a model inmate.

'How did you end up here in maximum security?' I asked.

'For the young guys,' he said. 'They [the prison psychologists] told me it would be good for me to participate in the therapy so that I don't fall back into any violent behaviour and that kicking the drug habit, as I did, was useful to show the other kids here that drugs are no good for anybody. So that's what I do. I tell them about my life and how I could have been so much better.'

Mike scored an 11 on the Psychopathy Checklist. It's a very low score for an inmate in maximum security, well below the average inmate score of 22 on the Psychopathy Checklist. Mike's not going back to prison, I figured; he's learned from his mistakes.

Mike suggested that I interview his cellmate. 'I think there is something wrong with him. Maybe you can figure it out, eh?'

I followed Mike back up to his cell and chatted up his roommate, 'Bob'. Bob was full of energy and enthusiasm, up for just about anything. Bob was a character. I could barely get a word in during our interview. I'd ask him a question and then next thing I know he's been telling me stories for fifteen minutes. It was the first interview where I laughed out loud. I wasn't exactly sure if someone in my position was allowed to laugh out loud when conducting a clinical interview, but I just rolled with it, egging him on and telling him that he was hilarious. And honestly, the things he'd got up to – well, they were funny.

Because I wasn't in an adversarial role with the inmates, they tended to be very open and forthcoming in our interviews. Everything they told me was confidential. It was not going to end up in their files or get them in any trouble. (And, as I mentioned earlier, I've changed the names and details so that no one reading this book can identify about whom I am talking.) I was there because I wanted to understand them. I remembered feeling like Columbo when I first started working in prison; I wanted to know what made these guys tick, how they got this way. My goal at that time was to use what I learned in my interviews to try to improve risk assessments.

In the mid-1990s in Canada, thousands of offenders went up for parole every year. Parole boards, composed of appointed individuals with widely varied backgrounds, had the power to decide who got out of prison and who stayed in. When an offender came up for parole, the board wanted to know what the chances were that the guy was going to reoffend. This could be a life-and-death decision for the inmate and for the public. It's certainly a decision with enormous economic and, of course, emotional consequences for society in general and for future victims in particular. Parole boards could interview the inmate and make their own release decision (never a good idea); they could ask a mental health professional such as a psychologist, social worker, or psychiatrist to make a professional judgment (again, not a good idea); or they could order a risk assessment workup be done on the offender and use that assessment to inform their decision (always a good idea).

Parole boards are typically not very good at predicting whom to let out of prison. A recent study showed that psychopathic offenders are more likely to convince parole boards to let them out compared to nonpsychopathic offenders.[2] This is an ongoing problem because we know that psychopathic offenders are more likely to reoffend than nonpsychopathic offenders. Professional judgment from a psychologist or psychiatrist has also been shown to be very unreliable in predicting who is going to reoffend.[3]

Because of the limitations of relying on guesswork by a handful of individuals or the unreliable opinion of one individual, scientists have turned to creating risk assessment tools or procedures that consider a number of variables to create an informed, detailed assessment of a specific offender. Criminologists and forensic psychologists have studied what variables promote risk and, similarly, what variables promote resistance to crime. Some of these variables are 'actuarial', like the age of the offender, gender of the victim, type of prior crimes, age of onset of criminal behaviour, and so on. These latter variables do a decent job at gross rankings of who is likely to reoffend and who is not, but they are not particularly good at predicting who will be a violent offender or who will commit a sex offense. Then there are psychological variables, like intelligence and personality, including psychopathy. Finally, there are dynamic variables, things that change, like marital status, education, employment, family relationships, and so on. These variables are entered into large databases, and scientists analyse them and try to determine which ones predict which inmate is going to get in trouble again. Armed with this information, researchers develop tests that parole boards can use to help make their tough decisions a little bit easier.

This was the landscape at the beginning of my career. Risk assessments were just coming onto the forensic scene, and my first goal was to try to improve the diagnostic sensitivity and specificity of predicting risk. I wanted to help the forensic decision makers identify who the high-risk offenders were, in particular the psychopaths, so we could make sure that they didn't get released, or at least not released without very special provisions in place so that they didn't reoffend.

The field of psychopathy research in the mid-1990s was still in its infancy. There had not been a single brain-scan study of psychopaths. It wasn't until 1991 that there was even a good way to do a clinical assessment of them (the year the first edition of the Psychopathy Checklist manual was published). The one thing that was known was that psychopaths were at very high risk to reoffend. An inmate who scored high on the Psychopathy Checklist was four to eight times more likely than an inmate who scored low to reoffend in the next five years – and more likely to reoffend violently.[4] Clearly, it was a critical component of any risk assessment.

Bob told me a story of how he used to hike across the border to the United States (there are few fences covering the three-thousand-mile border between Canada and the United States) and then hitch-hike down to the Indian reservations to buy cartons of cigarettes, as many as he could fit into his large backpack. He would hitch a ride back to near the border and hike back to Canada bringing with him dozens of cartons of cigarettes. Then he would sell them in Canada for a huge profit. Cigarettes were heavily taxed in Canada. He made quite a bit of money.

'Who do you sell them to?' I asked.

'Oh, that's the best part,' he answered. 'You have to sell them to people who won't turn you in; I sell them to pregnant women. They are too embarrassed to go into the stores to buy them, so they are always willing to buy them from you at more than they would even pay at retail in Canada. It's great. Easy money.'

Bob had done every type of criminal activity that I have ever heard of — and a few that I hadn't. He had done credit card scams, identity theft (I made sure my wallet was secure before I left the office), burglary, stolen cars, occasional mugging, random stints of drug use and dealing drugs and prescriptions (he once blackmailed a physician to write false prescriptions for him, running all over town to different pharmacies to get them filled), and most recently, Bob had been convicted of manslaughter.

Bob also had a fetish. He liked to be a Peeping Tom, and he collected women's underwear. After one arrest, the police found over three thousand pairs of women's underwear in his closet.

Bob described being questioned in his apartment on suspicion of burglary by the cops and sitting there in handcuffs on the couch.

'I warned the cop,' he said, 'not to open the closet.'

As the officer unlocked the door, the closet burst open and dozens of pairs of women's underwear landed all over the policeman (many of which were dirty; Bob seemed to prefer to steal them out of the laundry bags when he broke into a house — or from Laundromats).

I had tears in my eyes at this point from laughing so hard.

'Yup.' Bob laughed along with me. 'Even the cop's partner started laughing. What can I say? I love women's underwear. All kinds.'

'You know,' he said, 'can you tell me if I have ADHD? As a kid I was told that I had ADHD, but I don't know if I do or not. I mean, do you know how hard it is to hold still sitting in a tree to stare through the cracks in a Levolor blind on a second-floor window for three hours until someone comes into the bedroom and takes their clothes off?'

'Nope,' I said.

'Well, it's not easy,' he said, sitting back.

After another hour or so of stories, I was getting a pretty good picture of Bob.

'Okay, Bob. We've got to talk about your latest crime. What happened?'

'Oh, that. Well, it's pretty simple really. This girl I was living with, well, she pushed all my buttons. I mean, she hit all three, right in a row and I just got pissed. I ran after her into the bathroom where she was drawing a bath and pushed her really hard into the wall. She hit her head on the wall and slid into the tub, which was full of water. I just grabbed her around the throat and held her under the water. I was so pissed off. Then, ya know, bells go off inside my head ... oh, shit. I'm in trouble. Look what I did. I got to clean this up, figure out how to get out from under this crap. I mean, she was such a bitch to me that night.' Bob was animated, laughing at some of the 'funny' parts of this story. He used his hands a lot, gesturing about the entire sequence of events.

Across the table, I was feeling nauseated thinking about what he had done. I'd stopped laughing a while ago.

'So I wrapped her up in a big blanket, took her outside [it was dark], and put her in the car. Ya know, it was pretty stupid. I put her in the front passenger seat. Then I drove down the way a bit to a bridge and threw her over into the river, threw the blanket away in a Dumpster, and went out to create an alibi.'

'Where did you go?' I asked.

'I went and ate and drank some beer at my local pub, ya know, to act like nothing had happened. Then I went out and got a prostitute. I wanted to pay with a credit card, ya know, to get a receipt, but she wouldn't let me, so I had to go and slap her around a little bit, ya know, so she would remember me and stuff since she wouldn't give me a credit card receipt. It was nothing hard, ya know, but just enough so she would remember me, for my alibi.

'And then I went home. Went to bed. It was a couple days later when things started to unravel. Her mom kept calling, looking for her [he says this with a confused look on his face as if he doesn't understand why his girlfriend's mother would worry about not hearing from her daughter], and I told her mom that we had had an argument and she had packed a bag and moved out.

'The police came by and questioned me a bit. I just told them the same story, told them where I was the night she left. Ya know, I went out with a prostitute. Go check it out.

'They [the cops] kept coming back and forth to see me, but I never changed my story. After about a month, the mom was driving the cops crazy and they kept coming back to see me, would handcuff me, tell me they found the body [they were lying], all sorts of stuff. They even put a camera on me. I played a good trick on the cops, though. When they were recording my interview, I told them I wanted a lawyer, I repeated over and over. And then I could just say, later, ya know, they refused to give me my lawyer.

'I figured that they would find the body someday, and then I would be really screwed. So I figured if I confessed on tape after they refused to get me my lawyer, I might get the case tossed out on a technicality, and then they could never charge me if they found the body and stuff. So, well, I kept asking for a lawyer while the camera was on, and then I finally told them that I did it, that I killed her. But I told them to get me a lawyer. I didn't tell them where I dumped the body.

'So then they finally get me a lawyer. And I tell him the story about the videotape and failing to get me my lawyer and stuff. It's like against my rights, eh?

'Well, my lawyer tells me there is no tape the cops have given him, no record of any videotape. So, he says, you confessed. Now tell me where the body is and I will get you a deal. I got so pissed off, ya know? This shit works all the time on television. Anyways, the lawyer got me a deal, manslaughter, and I'll do a nickel or so and then get parole and it's all good. No worries.'

Another difference between psychopaths and other inmates is that psychopaths don't get distressed by being in prison. Most inmates get depressed when they get inside, and they find prison to be a stressful experience. A hallmark feature of psychopaths' disorder is that they don't get bothered by much of anything. They don't ruminate and they don't get depressed.

Bob scored 35 out of 40 on the checklist, a clear psychopath. I thought about telling Mike what was wrong with his cellmate, but that would break confidentiality. Mike would have to go on wondering.

Five years later, while I was still at RHC prison, Bob came bouncing up to me and said, 'Hey, still doing that research? I'd love to do that again.'

I stared at him. 'What are you back in for?'

'Oh,' he smiled and said, 'another chick pushed all my buttons.

What's a guy gonna do?' He laughed and walked away. Bob's buttons? His girlfriend had called him *fat, bald,* and *broke.* 'She hit all three of them,' he would tell me in his next interview, 'but I buried this body real good.'

After my first interview with Bob, I headed back to my office. I passed Grant on the way.

'Hey, Kent, things go okay today?' Grant asked.

'Just fine,' I replied.

'Good, glad to hear it, we guys [he's referring to the regular inmates] like you here. Just be careful around those sex offenders.'

'Sure will,' I answered. I opened the door and locked it behind me with my brass key and headed down to my office.

Something just wasn't right. I'd been too careful to piss off an inmate, especially one I hadn't even met yet, someone I hadn't even challenged in an interview.

I left the facility, taking a deep breath as I passed through the final gate and inhaled my freedom.

Day 3

After the normal morning commute, coffee distribution, and visit to the printer to pick up fresh copies of my evolving Psychopathy Checklist interview, I headed to the main door to psychiatry and the now-familiar corridor to the inmate housing units. As I shut the door, I noticed a figure at the end of the walkway. He was unmistakable – the large, ominous figure was Gary. He was just standing there, right inside the door from the housing pods. I knew that there were no cameras in this little area. I was worried that Gary knew that too. I turned and slowly locked the door with my brass key, trying to avoid him noticing that I was aware of him. I tried to remember any self-defence moves I knew, in case my nemesis attacked me when I reached the end of the corridor. My heart was racing.

Behind me a door opened and I let two inmates pass by. It took me a second to realize that one of the inmates was Grant. He turned, gave me a little wave hello, and kept going down towards the housing unit.

Gary had been staring at me. He was standing up straight to see past the other inmates, keeping his cold stare on me as the other inmates approached him. Grant slowed and took another peek back at me and then at Gary. Grant walked right up to Gary. Gary looked down at him, and a few words were spoken. Gary looked hard at Grant. All this was transpiring as I walked as slowly as possible, while still walking normally. It felt like my life was proceeding in slow motion. Then suddenly Gary turned, shot one more cold look in my direction, and went back into the housing unit. Grant followed him, without saying a single word to me.

A few hours later I was sitting alone in the nurses' station and Grant appeared at the half door.

'Hey,' he said, 'I wanted to let you know that I took care of that little problem with Gary. Seems he wanted to get it on with you this morning, but I told him you were not to be touched or it would be like he had touched me. And nobody fucks with me. I'll see you later.' With that, he disappeared down the housing unit hallway. I wasn't even able to choke out a word in reply.

A cold sweat broke out across my forehead. I might have actually been in a fight this morning with an enormous sex offender.

I left prison early that day and drove home in silence, deep in thought. I grabbed my dog, Jake, and went for a run along the beach to decompress. Later, over dinner, I finished reading *Games Criminals Play*. I had an epiphany.

Day 4

I took a new route to my prison office the next morning, avoiding the common areas where the inmates had free movement. I passed out coffee beans to a couple of other guard pods, receiving smiles and thank-yous. I was smiling too, but I was a little nervous about what I was about to do.

As soon as I was settled in my office, I called over to the nurses' station. Dorothy picked up.

'Do you have Grant down there this morning?' I asked.

'Hold on,' she said. 'Yes. He's still in his cell. Want me to get him for you?'

'Please. Send him down to psychiatry.'

'Got any more coffee?' she asked.

'Yup. I'll be right down after this quick interview.'

'He'll be right down then,' she said.

I went out to the main door and waited for Grant, dangling my little brass key.

Grant appeared a few minutes later and walked quickly up the corridor. He was carrying a folder with paper in it.

'You got more research for me to do?' he said.

'Yup,' I said. *Something like that,* I was really thinking.

'Good. Say, can you do me a favour this morning? I have this folder with my homework in it for group later today, and I need to make some photocopies so I can share it with the other guys.'

'You know you have to order the photocopies and they come out of your personal fund,' I replied.

'Yeah, but we're friends; I helped you out, ya know. I went to bat for you with that sex offender.' His voice turned a little coarse. 'You owe me.'

'Sit down,' I told him firmly. 'You and Gary are trying to scam me.'

He stared at me. 'That's bullshit. I don't work with sex offenders, man. This is the thanks I get for saving your life?'

There was some truth to his statement – violent offenders never associated with sex offenders.

I stared back at him and then pressed him again: 'I'm calling you out; you are trying to play me.'

And then he cracked. 'Is this conversation still confidential?' he asked.

'Yes,' I replied.

He fell into a full laugh. 'How'd you figure it out?'

'I'm new, but I'm not stupid,' I replied, not mentioning that I had lost quite a bit of sleep this week trying to work out what the hell was going on, or that I felt like I had aged a year in only a couple days.

'No hard feelings, right? Ya know, we have to test you. It's just something we do with all the new guys.' He was smiling ear to ear.

'I can tell ya that we've gotten quite a bit of fun out of a few folks before. Gary and I have been perfecting our moves; we thought we had you going pretty good,' he said. 'Especially since you told us everything we did with you was confidential. We figured unless we really hurt you, we were golden; you can't tell anybody about the scam.'

I now had a useful warning to pass along to any colleague working in prison — make sure your inmates don't use the confidentiality, which is there for the inmates' protection, to run a scam on you.

'What was the end game?' I asked. 'What were you hoping to get out of me?'

'Cigarettes, maybe some pot,' he said, shrugging, 'stuff like that. You wouldn't get in too much trouble if you got caught. We were just trying to enjoy life a little more, make time go a little faster.'

Something told me that Grant and Gary would not stop at a cigarette or other contraband.

'Well, I've got to get back to work,' I said. 'I'm glad we got this cleared up, and I'll keep it to myself,' I noted. I wasn't going to say anything to anybody; I was embarrassed I had been played.

'Okay. I'll tell Gary that the game is over. Seriously, no hard feelings, eh? I still want to do the research.'

'I'll be in touch,' I said. 'You are an excellent research subject,' I quipped as he stood up to leave my office.

'Oh, and take this to Dorothy,' I said, handing him a bag of coffee. 'If it doesn't get to her, then I know who to come looking for.' He laughed as I led him back into the corridor to the housing pods.

I sat down in my office chair and took a deep breath. I scribbled some notes on my pad, *Prison is never boring*, I wrote. The phrase became one of my favourite sayings to describe the environment in which I was going to spend the greater part of the next twenty years. Prison is never boring.

After reviewing the files of a few other inmates, my interviews for the day awaited. I proceeded down to the housing units to schedule a few inmates for interviews, and I decided to have a cup of coffee with Dorothy.

We chatted about the ins and outs of the RHC. I picked up details about when new inmates were coming in, the history of the therapists conducting treatment, the procedures to follow to keep the guards from getting angry with you. Essential information for someone rounding out his first week in prison.

Gary appeared at the end of his tier, his familiar stone cold face gone. A small smile crept across his face. He nodded slightly, turned around, headed back down to his cell.

'Back to work,' I told Dorothy.

'Back to work,' she echoed.

I headed down the sex offender tier for the first time. I passed by the cells until I read Gary's name on the cell door.

'Gary,' I said, peering into a dark cell. He had pictures of hockey and football players on his cell walls.

'Figured I'd hear from you at some point,' came the answer out of the darkness.

Gary emerged from the recesses of his cell and came to the door.

He was a big man. I couldn't help but notice the muscles bulging out from under his white T-shirt. I was thankful that we did not end up in an altercation.

'So you gonna do this research stuff or not?' I asked.

Laughing, he replied, 'Sure. That's all you want to say to me?'

'We can talk about stuff once you've signed all the consent forms and the confidentiality documents,' I said.

'So I don't have confidentiality yet,' he said in a nervous tone.

'Nope, not unless you sign up for research,' I said, realizing that I might be manipulating a research subject into participating. *Oh, well*, I thought, *the ethics board would understand.*

'Okay. Why not? Guys around here say you're fun to talk to,' he replied.

'Sign this form – it allows me to check out your files – then I'll come get you after lunch today for your interview.'

'No problem,' he said. 'But it might take you longer than that if you are going to read *all* of my file.' He laughed and sat down on his bed.

'Well, two p.m., I'll come get you.'

I walked out of the tier and straight down to the file room, requesting Gary's file.

Normally, the clerk would return with a folder or two, about two to four inches thick. The files contain police reports, institutional behaviour reports, social worker histories, interviews with other mental health personnel, family and work histories, school records, and such. In this case, the clerk returned with a six-inch-thick file, bound together with several thick rubber bands. *That's not so bad*, I thought, picking up the file. Then the clerk said, 'Hold on, that's just the first folder.' She called back, 'You want this in a box or do you want to try and carry it?'

'Umm, box, please,' I said. Gary's file filled a copypaper box. He was right; it would take me more than a lunch hour to read his file.

It turned out to be one of the most amazing case histories I would ever read. Gary would become my first perfect score – a 40 out of 40 on the Psychopathy Checklist, one of only a handful that I would find in the next twenty years.

Gary also became the very first psychopath to participate in brain scans a few years later.

After passing Gary and Grant's charade, the inmates accepted me into their circle. They signed up en masse to participate in research. During my seven-year tenure at RHC, over 95 percent of inmates volunteered for research. And so my career was under way, interviewing hundreds of inmates, cataloging their life histories, assessing their symptoms and personality traits, and eventually, studying their brains.

chapter 2

Suffering Souls

FACT: There are over 29,000,000 psychopaths worldwide.[1]

In 2008, author John Seabrook of *The New Yorker* **wrote** a feature article about my laboratory with the title 'Suffering Souls'.[2] As part of his research prior to writing the story, John visited my lab several times and had a tour of one of the New Mexico prisons where my team and I conduct research. At the outset, Seabrook didn't know the historical research that has been conducted on the psychopath, instead relying heavily on the way psychopaths are portrayed in popular film and media. Seabrook and I spent several weeks discussing the history of psychopathy, and he wove a wonderful story capturing the status and controversies of the field at the time.

As I told Seabrook, a little less than 1 percent of the general population, or about 1 in 150 people, will meet criteria for psychopathy. However, the number of psychopaths in prison is much higher than in the community because psychopaths tend to get themselves in trouble with the law. Studies indicate 15 to 35 percent of inmates worldwide will meet criteria for psychopathy – with more psychopaths being found in prisons with higher security ratings. I told Seabrook that something *special* seems to happen when the majority of psychopathic traits coalesce in the same person. I've often been quoted as saying that there is 'just something different' about psychopaths.

Nevertheless, it's important to recognize that psychopathic traits exist, more or less, in all of us. Fortunately, the distribution of psychopathic traits is skewed – *most* people have very low levels of the traits, *some* people have a bit more of the traits, and only a *few* people have high levels of the majority of the traits. It's the last group that scientists reserve for the diagnosis of *psychopath*.

But the term *psychopath* continues to be (mis)used in a wide variety of contexts. For example, it is not uncommon for the media to declare that a Wall Street trader, politician, or 'deadbeat dad' is a *psychopath*. In this context the media are typically using the label *psychopath* as a derogatory term, and they are not referring to the scientific definition of the disorder. When I am asked about such offending politicians, I typically reply that such individuals may be a bit *more* psychopathic than the rest of us, but I prefer to leave the diagnoses of *psychopath* for those few among us who have the full manifestation of the disorder.

Let's examine how history has recorded those among us who are 'just different' from the rest of us.

A Brief History of Psychopathy

Like Seabrook, I have borrowed the 'Suffering Souls' chapter title from German psychiatrist J.L.A. Koch (1841–1908), who is credited with coining the term *psychopathische,* or *psychopath.*[3] The term psychopathy literally means 'suffering soul'.

Psychopaths, under a different label or terminology, captivated attention long before Koch. Indeed, since humans first evolved, history has recorded stories of humans who display what we now understand to be the disorder psychopathy.

Perhaps the earliest written description of psychopathic traits can be found in the Book of Deuteronomy, from about 700 BCE.[4] About three hundred years later, one of Aristotle's students, Theophrastus (371–287 BCE), became the first scholar to write about psychopaths in any detail. He named his prototypical psychopath 'the Unscrupulous Man'.[5]

Stories of psychopaths pervade literature. Greek and Roman

mythology is strewn with descriptions of such characters. Accounts populate the Bible, beginning with Cain – the first murderer. Psychopaths have appeared in stories from all cultures: from Shakespeare, including Richard III and Aaron the Moor in *Titus Andronicus*, to the villain Ximen Qing in the seventeeth-century Chinese epic *Jin Ping Mei* (*The Plum in the Golden Vase*). More recently we have Macheath from Bertolt Brecht's *Threepenny Opera* and Hannibal Lecter from Thomas Harris's book *The Silence of the Lambs*.

Psychopaths also appear in existing preindustrial societies, suggesting they are not a cultural artefact of the demands of advancing civilization but have been with us since our emergence as a species. The Yorubas, a tribe indigenous to southwestern Nigeria, call their psychopaths *arana-kan*, which they describe as meaning 'a person who always goes his own way regardless of others, who is uncooperative, full of malice, and bullheaded'.[6] Inuits have a word, *kunlangeta*, that they use to describe someone whose 'mind knows what to do but he does not do it', and who repeatedly lies, steals, cheats, and rapes.[7]

Psychopaths are typically described in historical texts as monsters, evildoers, people who lack the emotional connections that bind the majority of us, as well as the inhibitions that those connections engage. Simply put, psychopaths lack conscience and empathy.

Psychiatrists, like the rest of us, have also been fascinated with psychopaths. Clinicians have written about psychopaths since the birth of psychiatry in the early 1800s.

History of Psychopathy in Psychiatry

The first psychiatrist to write about the condition was the director of the most famous insane asylum in the world, the Bicêtre Hospital, just outside Paris, France. Philippe Pinel (1745–1826), the founding father of modern psychiatry, described a group of patients afflicted with *mania sans délire* (insanity without delirium).[8] The term was used to describe cases of individuals who had no intellectual problems but a profound deficit in behaviour typified by marked cruelty, antisocial acts, alcohol and drug use, irresponsibility, and immorality.

What struck Pinel was that the intelligence of these individuals was generally very high, but something was not right with them. Pinel had been trained to view the mentally ill as only those who had delusions or distorted awareness. But as he discovered more and more of the latter patients who suffered from a severe deficit in morality, he changed his views. Pinel was one of the first to describe a type of *insanity* that did not occur with a commensurate confusion of mind and intellect, differentiating these cases from patients with psychotic behaviours.

It is important to understand that psychotic patients (i.e., psychosis) and psychopathy are very different. Psychosis is a fragmentation of the mind, leading to symptoms that include hallucinations, delusions, and disordered thoughts. A critically acclaimed film illustrating psychotic symptoms is *A Beautiful Mind*, in which actor Russell Crowe portrays Nobel Prize–winning mathematician John Nash. Nash suffered from hallucinations and delusions but was still able to develop game theory that revolutionized the field of economics. Psychosis is manifest in disorders like schizophrenia, bipolar disorder, and major depression. These psychotic symptoms are not typically observed in psychopaths.[9] Indeed, it was the absence of psychotic symptoms that originally differentiated psychopaths from other patients in mental hospitals.

Echoing Pinel, the American psychiatrist Benjamin Rush (1745–1813) argued that the moral faculty, like the faculty of intelligence, was susceptible to brain damage and should be included in the realm of medicine.[10]

As the United States and the United Kingdom developed their legal systems, the notion of free will and responsibility took centre stage as psychiatrists began to describe more and more mental illnesses that were associated with impairments that might lead to insanity defenses for criminal behaviours. From the beginning, psychiatry and law have had a tumultuous relationship. This tension was never more evident than in the development of the condition 'moral insanity' first proposed by James Prichard (1786–1848).[11] Prichard's use of the term *moral insanity* was broad and encompassed nearly all mental illnesses except for schizophrenia and mental retardation.

In the 1800s, the concept of moral insanity led to a lively debate in psychiatry on both sides of the pond. Leading British psychiatrist Henry Maudsley (1835–1918), a supporter of Prichard, stated in 1875 that:

As there are persons who cannot distinguish certain colours, having what is called colour blindness, and others who, having no ear for music, cannot distinguish one tune from another; so there are some few who are congenitally deprived of moral sense. (p. 58)[12]

Prominent supporters of moral insanity, such as Isaac Ray (1807–1881), argued that the emotional and intellectual psyches were of equal importance with respect to the definition of *insanity*. Ray believed that behaviour emanated solely from the brain and that aberrations of emotional powers were connected with abnormal conditions of the brain.[13]

The concept of moral insanity and its subsequent related brethren thus set out on a long and complicated history with the legal system. On the one hand, psychiatrists argued that patients afflicted with psychopathic symptoms were as disordered as patients with schizophrenia; on the other hand, many people had serious concerns with what was essentially the medicalization of criminal behaviour.

Moral insanity continued to be debated in legal and academic circles as books and other publications proliferated arguing one side or the other. In part because of the controversy of the term *moral insanity* and because of the generality of the term as it applied to many who were simply criminal, German psychiatrist J.L.A. Koch (1841–1908) coined the term *psychopastiche*, or *psychopath*, in 1888.

Koch was also one of the first to insist that to assess psychopathic traits one has to consider the entire life history and all facets of an individual's life in order to get a complete picture of how the symptoms manifest themselves. This critical piece of the puzzle has shaped much of how we currently assess psychopathic traits in forensic settings. Koch's diagnostic criteria were very popular and were included in the eighth edition of Emil Kraepelin's classic textbook on clinical psychiatry.[14]

Even though Koch's *psychopastiche* narrowed the term from the vast population of those criminals with moral insanity, it was still a broad construct that was indicative of a generic label for all personality disorders. The so-called German School of psychopathy expanded the category to include people who hurt themselves as well as others, and in the process of broadening the definition seemed to lose sight of the moral disability that was central to the condition.

By the 1920s, psychiatry was using the word *psychopath* to include people who were depressed, weak willed, excessively shy, and insecure – in other words, almost anyone deemed abnormal. For example, in one of my favourite films, *Changeling*, a tragic true story set in 1928 Los Angeles, Christine Collins (played by Angelina Jolie) is incarcerated in the 'psychopathic ward' for her emotionality associated with repetitive challenging of the police's treatment of the case of her missing son.

Sociopathy Versus Psychopathy

Up to this point in time, the construct of psychopathy had been grounded in Koch's physical or biological basis. This view soon collided with the emerging behaviourist era in psychology. Behaviourists believed the human brain was a blank slate at birth and all processes, even psychopathic traits, were formed through social forces.[15] The term *sociopathy* was introduced in the 1930s and signified the origin of psychopathic traits from social causes. One of the problems with the diagnosis of sociopathy was that it was too broad and encompassed far too many individuals. Literally every criminal met criteria for the condition. Another unfortunate consequence has been that the term *sociopathy* has been confused with *psychopathy* ever since.

Sociopathy and psychopathy are very different. Sociopathy includes a broad, heterogeneous category of individuals who act antisocially, the causes of which are believed to be social and environmental in nature. *Psychopathy* is a term grounded in biology and genetics and is truly agnostic to causes or etiology. In other words, genetics and the make-up of the brain, as well as environment, contribute to the construct of psychopathy. Although the term *sociopathy* is *not* used in modern academic circles to mean 'psychopathy' anymore, some people continue to confuse the terms.

Psychopathy Assessment Research

The first formal meeting on psychopathy was organized by forensic psychiatrist Ben Karpman (1886–1962) and held at St Elizabeths Hospital in Washington in 1923. Karpman felt that because of psychologists' general lack of knowledge of the field, the question of psychopathy's origins (nature vs. nurture) should be held open until the exact symptoms of the condition were clarified. German, English, Italian, and American scientists came together at the meeting to develop the first comprehensive set of symptoms to characterize psychopathy.[16] One central theme was that psychiatrists were constantly surprised by the behaviour of psychopaths. Psychopaths were smart, and it baffled psychiatrists that they are unable to use that intelligence to control their behaviour.

I often refer to psychopaths as a *walking oxymoron*: they say one thing and do another. Clinicians in the early days of psychiatry were mystified, confused, even stymied by the repetitive self-defeating behaviour of psychopaths once they left the confines of the psychiatrist's office, mental ward, or prison cell. Psychiatrists felt that psychopaths should simply learn from their bad outcomes (divorce, bankruptcy, arrests, conflicts with parents, siblings, and friends) and not repeat their poor decision-making. But, to the chagrin of those conducting psychotherapy, psychopaths rarely if ever changed their pattern of behaviour once they left the psychiatrist's sofa.

Twentieth Century

The psychoanalytic school of thought, founded by Sigmund Freud (1856–1939), dominated much of the early twentieth century. Freud argued the psyche could be divided into the id, the ego, and the superego. The superego is supposed to develop as a result of the Oedipal conflict – a conflict that arises in a boy's unconscious desire to sexually possess his mother and kill his father. A normal superego puts the brakes on impulses and aberrant sexual behaviour. If the psyche is damaged or poorly developed, it lacks the ability to suppress the id (instinctual drives), and antisocial behaviour is the result.

Psychodynamic thinkers wrote that psychoanalytic treatment of psychopaths was never successful. Indeed, the psychopath's ego was fed by the therapist's interest in him. Thus, while trying to treat the psychopath, many psychodynamic therapists found the psychopath only got worse and more egocentric.

This belief carries through to popular culture today. For example, after years of performing psychotherapy on the fictional mafia figure Tony Soprano (portrayed by James Gandolfini) on the hit American television programme *The Sopranos*, his therapist declared that she had finally worked out that Tony was an incurable psychopath and so she quit treating him.

Psychopaths are resistant to psychodynamic treatment, in part because they typically don't feel there is anything wrong with them; they are rarely interested in participating in therapy or changing. Consequently, psychodynamic therapy was not very successful with psychopaths, and that led to the pervasive view that persists today that psychopaths are untreatable.[17]

Freud's psychodynamic theory in general and psychoanalytic treatment in particular quickly fell under heavy criticism when it was pointed out that it was not a theory at all; that is, psychodynamic theory does not make any predictions, and hence does not have any testable hypotheses.

It was American psychiatrist Hervey Cleckley (1903–1984) who changed the face of psychopathy research with his masterpiece *The Mask of Sanity*, first published in 1941.[18] Cleckley sought to clarify the construct of psychopathy from other incantations of the disorder and to provide the psychiatric community with a treatise on the condition using a plethora of case studies. From these case studies, Cleckley gleaned sixteen criteria that formed the basis for the modern assessment of the condition. In elegant, magnificent prose, Cleckley educates the reader about the symptoms and make-up of psychopathy. Cleckley's criteria, listed in Box 1, are consistent with the historical traits identified by his predecessors but contain much more detailed analyses and refinement. In four subsequent editions of *The Mask of Sanity*, covering nearly fifty years of clinical experience, Cleckley continued to refine and illustrate psychopathy as we know it today.

BOX 1

The 16 characteristics of the psychopath as chronicled in the five editions of *The Mask of Sanity*, by psychiatrist Hervey Cleckley.

1. Superficial Charm and Good 'Intelligence'
2. Absence of Delusions and Other Signs of Irrational Thinking
3. Absence of 'Nervousness' or Psychoneurotic Manifestations
4. Unreliability
5. Untruthfulness and Insincerity
6. Lack of Remorse or Shame
7. Inadequately Motivated Antisocial Behaviour
8. Poor Judgement and Failure to Learn from Experience
9. Pathologic Egocentricity and Incapacity for Love
10. General Poverty in Major Affective Reactions
11. Specific Loss of Insight
12. Unresponsiveness in General Interpersonal Relations
13. Fantastic and Uninvited Behaviour with Drink and Sometimes Without
14. Suicide Rarely Carried Out
15. Sex Life Impersonal, Trivial, and Poorly Integrated
16. Failure to Follow any Life Plan

It is impossible to overstate the value of Cleckley's contribution to the field of psychopathy. The multiple editions of *The Mask of*

Sanity revolutionized our understanding and classification of the disorder.

The Hare Psychopathy Checklist

Beginning in the mid-1970s, some clinicians began to rethink a working clinical definition of *psychopathy*. Based on Cleckley's published criteria, Dr Robert D. Hare (1934–) and his students developed a clinical rating scale for the assessment of psychopathy.

The first version of the Psychopathy Checklist (PCL) was published in 1980.[19] In 1991, the PCL was revised by consolidating a few items and updating and optimizing the scoring criteria.[20] Since the publication of the Psychopathy Checklist-Revised in 1991, the numbers of items and scoring criteria have not changed. The manual was revised again in 2003, indexing the latest research that had been conducted with the instrument in the prior decade.[21]

For the first 150 years of studying psychopaths, experts did not have a common diagnostic instrument to assess individuals. But since the development of the Hare Psychopathy Checklist-Revised, we have had more than twenty years of research using the same instrument, and a very large literature has developed in the scientific community confirming its reliability and validity. Many studies have confirmed that psychopathy can be reliably assessed with the Hare Psychopathy Checklist-Revised. However, it should be noted that in the context of an adversarial proceeding in the legal system, the Psychopathy Checklist, like any other psychological instrument, can be distorted. To avoid such problems, the judicial system should avail itself of experts who are hired by the court, and not by one side or the other, to resolve the problems. It is this literature that forms the scientific backbone for the remainder of this book (see Figure 1).

For most clinicians and researchers, the Hare Psychopathy Checklist-Revised (henceforth referred to as the Psychopathy Checklist) has become the 'gold standard' diagnostic tool for psychopathy. It is the single most commonly used tool to assess psychopathy in general, and forensic populations in particular. It has been trans-

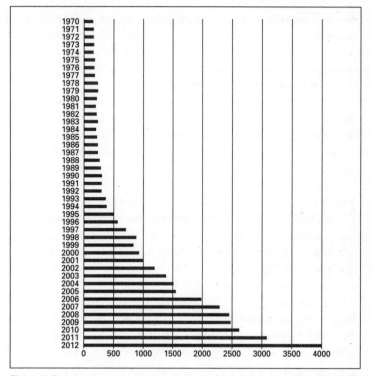

Figure 1. Google scholar results for number of published papers per year from 1970 to 2012 in the scientific literature with the keyword search 'psychopathy'. There is a massive burst of activity beginning about 1993– an effect likely due to the development of the assessment for psychopathy known as the Hare Psychopathy Checklist-Revised (PCL-R), published in 1991. (From Google scholar search completed 12/7/11; years 2011 and 2012 added on 6/11/13).

lated into sixteen different languages and distributed around the world.

As outlined in Box 2, the Psychopathy Checklist includes nearly all the traits that Cleckley and his predecessors developed, as well as additional traits that assess socially deviant behaviour.

To complete the Psychopathy Checklist, a trained clinician needs to do both a detailed semistructured interview of the client that lasts approximately two hours and a collateral file review. Files typically include police reports, social worker assessments, details on family dynamics, employment history, education, relationships, childhood, adolescence, and criminal history.[22] From these sources of information, the interview and files, trained experts then rate the client on the twenty items, following the scoring criteria in the Psychopathy Checklist manual.

I've trained hundreds of people to score the Psychopathy Checklist, including undergraduates, clinical and research graduate students, psychiatric residents, and psychiatrists and psychologists.[23] With about a week of intensive training and scoring of practice cases, nearly all the trainees are able to achieve reliable scores on the Psychopathy Checklist.

Dr Robert D. Hare is now known as the modern father of the field of psychopathy. In 2010, he was awarded the Order of Canada, one of the highest civilian honours bestowed by the country, for his scientific and community efforts developing the Psychopathy Checklist.

BOX 2

The 20 items of the Hare Psychopathy Checklist-Revised are a clinical rating instrument and the gold standard for the assessment of psychopathy since 1991.

Psychopathy Checklist Items

1. Glibness/Superficial Charm
2. Grandiose Sense of Self-Worth
3. Need for Stimulation
4. Pathological Lying
5. Conning/Manipulation
6. Lack of Remorse or Guilt
7. Shallow Affect
8. Callous/Lack of Empathy
9. Parasitic Lifestyle
10. Poor Behavioural Controls
11. Promiscuous Sexual Behaviour
12. Early Behavioural Problems
13. Lack of Realistic, Long-Term Goals
14. Impulsivity

15. Irresponsibility
16. Failure to Accept Responsibility for Own Actions
17. Many Short-Term Marital Relationships
18. Juvenile Delinquency
19. Revocation of Conditional Release
20. Criminal Versatility

Psychopathy and the *Diagnostic and Statistical Manual (DSM) of Mental Disorders*

While almost everyone recognized the importance of the affective traits Cleckley and Hare articulated, some psychiatrists had doubts about average clinicians' abilities to reliably detect affective criteria, such as lack of empathy, guilt, or remorse.

One common mistake that leads to overrating some of the affective items, such as *Lack of Empathy*, is to focus on a single bad thing that the person did, such as the index crime the individual committed that prompted his or her assessment and scoring. For example, if an individual commits a sex offence against a child, many trainees in the room will raise their hands when asked if this behaviour merits a high score on *Lack of Empathy*. But they are wrong.[24] The sex offence is only one piece of evidence suggesting impairment in empathy. To score high on *Lack of Empathy*, an individual must have evidence of the trait from multiple life domains and for the majority of his or her life – echoing Koch's seminal contribution to the assessment of personality (disordered) traits. So the individual who committed a sex offence against a child may very well deserve a high score on *Lack of Empathy* but if they do, it will be because the person has demonstrated impaired empathy for a long time in other areas of his or her life as well – the single offence alone does not automatically warrant the high score.

One trick we teach clinicians when rating items on the Psychopathy Checklist is to ignore the index offence – the offence for which a client is convicted or incarcerated. The individual should get the same psychopathy item scores regardless of the crime that leads to his or her imprisonment. This avoids the common issue where one monstrous deed leads raters to score the individual high on all traits.

Without proper training, the average clinician will likely have trouble producing valid ratings of psychopathy. The simple fix to this problem is that clinicians who need to perform psychopathy assessments as part of their practices or jobs should participate in a special professional training session. This is one of the reasons why continuing education is a required part of being licensed for any practitioner in psychology or psychiatry.

However, it was this tension – between those who did and did not think the affective traits could be reliably diagnosed – that drove the swinging pendulum of the American Psychiatric Association's *Diagnostic and Statistical Manual of Mental Disorders (DSM)* classification of psychopathy over successive iterations.

The *DSM* is the 'bible' of mental illnesses in the United States and influences doctors worldwide. It provides a template for how clinicians assess and classify patients into various categories of mental illness. Determining a patient's diagnosis is usually the first step towards determining the best course of treatment. However, defining *mental illness* is a complicated process, and I always teach my students that they need to go beyond the *DSM*. That is, the *DSM* is a good starting point, but if you really want to be on the cutting edge of the science of mental illness or developing new treatments, you must educate yourself about the strengths and weaknesses of any psychiatric diagnosis. Moreover, the *DSM* is an evolving document that is shaped by science, economics, and politics (not necessarily in that order). Thus, it is incumbent upon researchers that they understand the history of the mental illness they are studying and learn how the illness has been assessed in previous iterations of the *DSM*.

There was widespread dissatisfaction[25, 26] with early versions of the *DSM* treatment of antisocial personality disorder/psychopathy. This led the American Psychiatric Association to conduct field studies in an effort to improve the coverage of the traditional symptoms of psychopathy. As a result, *DSM-IV* (and *DSM-5*) reintroduced some of the affective criteria that *DSM-III* left out, but in a compromise, they provided virtually no guidance about how to integrate the symptoms.

Perhaps more important – and dangerous – forensic practitioners may diagnose a client with *DSM* antisocial personality disorder but mistakenly draw upon the literature using the Psychopathy Checklist and incorrectly relate the latter findings on recidivism and treatment outcomes to a client with only minimal antisocial symptoms.

I routinely consult with lawyers and judges, and it is not uncommon for them to have been told by their consulting forensic practitioners that the Psychopathy Checklist and *DSM* antisocial personality disorder criteria are the same when they are absolutely not.

In professional training seminars, I sum up the relationship between *DSM-IV* antisocial personality disorder and psychopathy as assessed with the Psychopathy Checklist this way:

The *DSM* antisocial personality disorder criteria get you about halfway to the diagnosis of psychopathy using the Hare Psychopathy Checklist-Revised. If you are a clinician working in the community and you complete an interview with your client and he or she meets *DSM* criteria for antisocial personality disorder, you know you are dealing with someone with a difficult personality. But then clinicians need to go beyond the *DSM* criteria and assess for psychopathy using the Hare Psychopathy Checklist-Revised. In this way the clinician will know whether he or she is dealing with a psychopath or not, drawing upon all the thousands of papers that have been published on psychopathy to help develop a treatment and management strategy for a client. If you are working in a forensic setting, you should simply skip the *DSM* criteria and use the Hare Psychopathy Checklist-Revised.

The evolution of psychopathy has been full of twists and turns, but the scientific community has finally sorted out a common metric for the condition. In the next chapter, I'll explore the symptoms of psychopathy in more detail, illustrated by case examples of two notorious assassins.

chapter 3

The Assassins

FACT: psychopaths kill more people in
North America every year than the number
killed in the 9/11 terrorist attacks.[1]

Twice in the latter half of the nineteenth century, the
world was stunned the assassination of a US president. The first assassination, of President Abraham Lincoln by John Wilkes Booth,
occurred in 1865. Sixteen years later, in 1881, Charles Julius Guiteau
assassinated President James A. Garfield. The two assassins used
the same instrument to execute their malfeasance, a .44-calibre re-
volver, but that is where the similarities end.

John Wilkes Booth was a prominent stage actor who led a relatively charmed life. A Southern sympathizer, Booth murdered Lincoln just days following the surrender of Confederate general Robert
E. Lee in an attempt to turn the tide of the Civil War back on the
side of the South. Charles Guiteau's crime, contrary to Booth's, was
utterly motiveless and simply mystified people. The trial of Guiteau for the assassination of President Garfield would fracture the
American medical and legal community over the diagnosis of *moral
insanity* and its relationship to criminal responsibility.

Many people assume horrific crimes indicate the perpetrator has
a disturbed, even deranged, personality. I am often asked whether
all murderers are psychopaths. Many people assume they are. But as
we've seen, psychopathy is more complicated than the details of any
single crime can capture, no matter how despicable the act. In the

pages that follow, I intend to assess the traits of psychopathy in two of the most notorious American assassins in history, to show how the scoring of an individual on the basis of those traits leads to a clinical diagnosis. As we know from Chapter 2, psychopathy, as assessed by the Hare Psychopathy Checklist, is comprised of 20 items that typify the affective, impulsive, and antisocial symptoms of the condition. Each of the items is scored on a three-point scale. A '0' indicates that the trait does not characterize the individual, a '1' indicates that the trait describes the individual in some areas, and a '2' indicates that the trait is present in all aspects of the individual's life. Typically, I would interview the individuals before scoring them, but of course Booth and Guiteau are deceased, so we will have to rely on the historical record. Thankfully, there is abundant information available on both men, due to the infamy of their crimes. We have available to us dozens of reports from family, friends, investigative journalists and reporters, autobiographies, historical biographies, and even personal diaries. These collateral sources are sufficient to score the items on the Psychopathy Checklist for both assassins. In scoring Booth and Guiteau, we have to be careful not to let our ratings of psychopathic traits be overly influenced by the assassinations both men carried out. Despite the fact both men committed a horrific crime against the nation, we cannot let a single incident distort our assessment; we must consider the totality of both men's lives.

My goal in this chapter is to offer readers a window into how scientists assess psychopathic traits in an individual. Typically, the Psychopathy Checklist is part of the risk assessment of criminal offenders, but the checklist is also being used in other parts of the legal system, including custody disputes, divorce court, and other adversarial processes. The assessment of these traits is not as simple as it might seem, and the results that follow might be surprising. Let's go through the traits, one by one, and assess John Wilkes Booth and Charles Guiteau in light of what we know.

1. Glibness/Superficial Charm

One aspect of psychopaths' behavioural repertoire is that they often speak quickly, volubly, and interrupt the flow of the conversation frequently, in an energized speech that observers can find difficult to follow and process in real time. The listeners are bombarded with so much information that they often leave the conversation not having been able to digest it all. Then, as observers recall the conversation, their minds interpolate, usually in a very positive sense, the information that was presented. The psychopath often comes off as quick witted, even likable, but the listeners' 'gut' feelings detect that there is something not quite right about the individual. It takes practice to sift through psychopathic speak.

One of my favourite things to do with university psychology undergraduates is to send them to prison to interview psychopathic inmates without letting them read the collateral files first or letting them know the individuals they are interviewing are psychopaths. I'll observe an interview and let the novice probe and question the subject. Upon completing the interview, I ask the student what he or she thought of the guy. More often than not I get a response such as 'He was so nice, I can't imagine why he is in prison' or 'If that guy was on the outside, I'd get a beer with him.' Then I let them read the inmate's file. 'This can't be the same guy,' a novice commonly replies. I tell the student, 'Go reinterview the inmate now that you have studied the collateral information.' During the reinterview, the novice asks, 'Why didn't you tell me about all that stuff in your file, all the crimes you committed, the rape, the robbery?' The psychopath more often than not replies with something like 'Oh, that's the old me. I wanted to talk about the new me.'

It's a valuable lesson for anyone who wants to work with psychopaths. It is not possible to score psychopathic traits based on an interview alone. You need collateral information if you want to score individuals with these traits.

Do you think that the victims of the psychopaths thought they were in danger? Of course not. They might have seen warning signs, they might have felt a little tug in their gut that something was off, but most victims of psychopaths don't understand the kind of person they are dealing with, or they would have stayed far away. As a psychologist, you need to trust that gut feeling and investigate. Don't be caught off guard. Being careful and prepared is never a bad thing when you work with psychopaths.

With regard to Charles Guiteau, evidence from business partners, his former wife, parents and siblings, friends, acquaintances, personal lawyers, and newspaper reports about his life and trial suggest he exhibited superficial glibness in all facets of his life. Guiteau would engage in conversations with such force and excitement that it would make an impression upon everyone with whom he came in contact. Guiteau earns a high score, a 2, on this item.

In contrast to Guiteau, John Wilkes Booth has been referred to as the Brad Pitt of his times. He was attractive, athletic, and an engaging stage performer from a family of famous thespians. He performed Shakespeare and other classics in over thirty cities. He helped introduce high-powered spotlights with coloured light to the stage to rave reviews and popularity. He was known as a ladies' man. Women, enthralled with his performances, literally lined up to go back to his room with him. Booth was considered genuinely charming and he had a sparkling personality. There is no evidence of psychopathology with regard to this trait, but he did embody a celebrity lifestyle, including an attitude of superiority, which merits some consideration. I score him a 1 (moderate) on our three-point scale (0, 1, or 2).

To sum up the scoring of the two assassins on this trait:

> Booth 1
> Guiteau 2

2. Grandiose Sense of Self-Worth

John Wilkes Booth aspired to follow in the footsteps of his famous father, Junius Brutus Booth. Junius was an international star, considered one of the foremost tragedians of British theatre before relocating to the United States in 1821. He toured throughout the United States to enormous acclaim. Poet Walt Whitman described him as 'the grandest histrion of modern times'.

John Wilkes's older brother, Edwin, also achieved significant popularity as a Shakespearean actor. Edwin is considered one of the greatest American actors to ever play the role of Hamlet.

John Wilkes got off to a rather stilted beginning in terms of his acting career. His initial forays were plagued by stage fright, an inability to recollect his lines, and clumsiness. John Wilkes Booth originally used his middle name as his last name in order not to tarnish his family's name as he tried to get his feet under him onstage and not to rely on his father's and brother's fame to jump-start his career.

Booth persevered and eventually became a star actor. He reportedly made over $20,000 a year in the 1860s, roughly $500,000 in today's currency, which placed him in the top 1 percent of all wage earners of his day. It was common back then for audiences to throw fruit, drinks, and even rocks at actors when they made mistakes. In some venues, it was not uncommon for rowdy members of the audience to fire guns in the general direction of a flailing thespian. Indeed, Edwin Booth describes in his autobiography such occurrences while he was touring California during the Gold Rush. The mood of the crowd could quickly turn sour if the entertainment their hard-earned money paid for was lacklustre. John Wilkes Booth learned to adapt and eventually thrive in this environment.

John Wilkes was something of a playboy. He described himself as arrogant. And we have to consider whether part of his motivation for murdering President Lincoln was an attempt to gain even more fame, as some historians have suggested. But most historians believe it was Booth's entrenched political beliefs that drove him to shoot Abraham Lincoln. He saw himself as a Confederate soldier, deeply dedicated to his cause, and believed that Lincoln was a tyrant who must be executed. He believed that the death of the president could be used to leverage the resurgence of the Confederacy. The only recognition he seemed to seek from his political crime, which he perceived as an act of war, was the approbation of his fellow countrymen. Moreover, John Wilkes does not appear to have an inflated sense of his self-worth. He was able to achieve, through hard work and practice, a sizeable following in the theatre. He maintained a close group of friends and family. His score on this item is moderate, at best a 1.

Guiteau, on the other hand, felt that menial work was beneath him. His former wife reported that he looked down upon people who did such work. He believed he should be a published author, but he failed to put in the time and effort in his writing, instead plagiarizing text from others. As an orator, he enjoyed the crowd's attention but copied his religious speeches from others. He dreamed of marrying rich, travelling the world, and living the high life, even though he had no reasonable plans to acquire such wealth. His former wife noted that:

> *He was always anxious to live so far beyond his means. It was always 'Nothing but the best', the best place and among the very best first-class people, prominent people, people well known, so far as position and wealth were concerned. That was his great object – always to be among them and to live at the most expensive places and to have the best accommodations; he was not satisfied to live in plain style anywhere. (pp. 85–86)[2]*

A note found in his pocket when he was arrested for shooting President Garfield provides additional evidence of this trait:

> *To the White House,*
>
> *The President's tragic death was a sad necessity, but it will unite the Republican Party and save the Republic. Life is a flimsy dream, and it matters little when one goes. A human life is of small value. During the war thousands of brave boys went down without a tear. I presume the President was a Christian and that he will be happier in Paradise than here. It will be no worse for Mrs Garfield, dear soul, to part with her husband this way than by natural death. He is liable to go at any time any way. I had no ill will toward the President. His death was a political necessity. I am a lawyer, a theologian, and a politician. I*

am a Stalward of the Stalwarts. I was with General Grant and
the rest of our men in New-York during the canvass. I have some
papers for the press, which I shall leave with Byron Andrews and
his co-journalists at No. 1,420 New-York Avenue, where the re-
porters can see them. I am going to jail. CHARLES GUITEAU.

A search of Guiteau's home following his arrest revealed numerous letters; one letter was to Vice President Chester A. Arthur (who
succeeded President Garfield upon his death) in which Guiteau
made recommendations for selecting members of his new cabinet.

A morbid follower offered to pay $1,000 for Guiteau's body fol-
lowing his execution. Guiteau replied, 'I think I ought to bring more
than that. . . . Perhaps some other fellow will offer $2000, then I can
pay my debts, and if I get a new trial, that miserable Corkhill can't
bring on a lot of fellows just to swear how much I owe them.' The
prosecutor, Corkhill, had subpoenaed numerous individuals who testified that Guiteau owed them money.

During his time in jail, Guiteau dictated his autobiography. The
man who took the notes in shorthand referred to Guiteau's vanity as
'literally nauseating'.

And, of course, Guiteau wanted to represent himself at trial. In
my clinical experience with prisoners, many psychopaths are so
grandiose that they feel they can do a better job defending themselves than an experienced lawyer could, despite the fact that most
of them didn't finish secondary school.

This egotism and arrogance is palpable in every aspect of Guiteau's life. He earns a perfect score on this item.

> Booth 1
> Guiteau 2

3. Need for Stimulation/Proneness to Boredom

This trait reflects an individual's need for excitement and change.
Psychopaths find many tasks boring and switch plans frequently.

The need for high levels of stimulation leads to poor academic, work, and relationship outcomes. However, it is not for lack of intelligence, but because they are so easily distracted.

John Wilkes Booth was reported to be an average if not slightly below-average student. He was, however, according to his siblings and primary-school teachers, a hard worker who, once he tackled a problem, was able to always repeat the process, even if schoolwork did not come easily for him. Booth, like most of his siblings, attended and completed boarding school.

Acting, too, did not come easily to him, but he maintained the same strong work ethic in his thespian career as he had in school. It is this dedication that argues against a high score on this trait. Indeed, in the theatre, Booth played the same roles over and over again, something that would be very difficult for individuals with high levels of this trait.

Booth, like his father, was a drinker. Occasionally, he would imbibe to excess and arrive late or miss performances. And he was a ladies' man. He had many opportunities to settle down but preferred the company of many different women, even prostitutes. So he does show evidence of moderate levels of this trait, although not with his family, employees, or teachers.

Guiteau failed to complete university and never really held on to a job. He felt that most jobs were beneath him. He wanted to live an exciting life, full of travel, status, and fame, but he refused to put in the work to earn it. He wrote to generals, senators, even presidents, requesting that he be posted to various ambassadorships or higher-level positions abroad. His need for excitement was a constant drive.

Individuals with this trait often dabble in drugs and use alcohol to excess. Guiteau's wife reported that he never used drugs or alcohol. But this conflicts with reports that he was arrested for public drunkenness and frequented bars. One needs to be very sensitive to conflicting information from different sources. Here, the records documenting Guiteau's arrests, and numerous other witness accounts to his drinking, are more credible than his wife's assertions that her husband did not imbibe.

Except for the lack of drug use, Guiteau is a classic portrait of this trait.

Booth 1
Guiteau 2

4. Pathological Lying

My favourite term for describing this item is *mendacity*. Nearly everyone lies at some point in life. We're not concerned with the white lie or social lie here.

Mendacity refers to pathological lying, often for no reason at all, even when facts can be readily checked. Historically, this item was designed to capture the interpersonal style of the individual. Individuals who score high on this trait often lie for no reason whatsoever and, when caught in a lie, are unfazed and unconcerned and just move on to the next question during the interview. As with all traits, this type of interaction must be present in most domains of the person's life, not just, for example, in business dealings.

With Booth we see no evidence of pathological lying. In fact, even in his numerous dealings with women, he was apparently very straightforward. He told them that he was not interested in marriage, that he was just interested in carnal pleasure and their company for a brief period of time. Booth's original plan, to be carried out with other Confederate sympathizers, was to kidnap Lincoln and force the North into a prisoner swap, to release the hundreds of captured Southern soldiers. According to his diary, Booth was motivated by stories of Southern soldiers being tortured in Northern prison camps to reveal the whereabouts of Confederate sympathizers hiding in the North. In planning the attempt to kidnap Lincoln, Booth and his conspirators lied to family and friends, but this type of lying is limited to a specific situation and has a purpose. It is not the type of pointless lying that characterizes this trait. Booth was not a pathological liar.

With Guiteau, we see clear evidence of mendacity, with his siblings and father growing up, with his former wife, and of course in the shady business dealings he was involved in throughout his life. His ex-wife reported during her interview following his arrest that it was impossible to believe anything Guiteau said unless one knew it to be a fact from other information.

Guiteau lied about his education, claiming he finished a bachelor's degree at the University of Michigan, though he attended only a few lectures. He lied about his occupation, often telling people he was a published author (he plagiarized his entire book); he advertised himself as an accomplished lawyer (he was disbarred); and he wrote that he was a devoted husband (his wife presented his mistress at their divorce trial). Here is how I score Guiteau and Booth on this trait:

Booth 0
Guiteau 2

5. Conning/Manipulation

This item is designed to assess a person's willingness to engage in all types of manipulation for personal gain at the cost of others (e.g., frauds and con jobs).

Guiteau manipulated everyone he came into contact with. His various swindles and schemes left hundreds of victims. He was disbarred in Illinois for fraud over a business deal, and he extorted family, friends, and business partners. He deliberately made friends just so that he could beg, borrow, or extort money from them. To support his vices, he manipulated his wife into working. He faked being president of a national bank to try to secure a loan to buy a newspaper where he could publish his musings and (plagiarized) works. After he fraudulently published his book *The Life of Christ* using a stolen publisher's logo (D. Lockwood & Co.), he used this false imprint to convince the publisher Wright and Porter to continue production of his book – but never paid them for the work. One other con was that he advertised himself as an attorney of law working in the Massachusetts House of Representatives but held no licence in the state (and had been disbarred in Illinois).

In Booth, we see no evidence of conning or manipulation. Prior to the assassination, Booth was known as a decent businessman, fair to his family and friends. He and his brothers donated to charity. Indeed, a statue of William Shakespeare, paid for by revenue generated by the Booth brothers' charity performances, still stands in New York's Central Park just south of the Promenade.

> Booth 0
> Guiteau 2

6. Lack of Remorse or Guilt

During cross-examination in the trial for the assassination of President Garfield, the prosecutor asked Guiteau: 'You have never hinted at any remorse?'

Guiteau answered: 'My mind is a perfect blank on that subject.'

Prosecutor: 'Do you feel any more remorse about rendering [President Garfield's] wife a widow and her children fatherless?'

Guiteau answered: 'I have no conception of it as murder or killing.'

Guiteau went on to offer a weak response that he 'regretted' the necessity of the crime but that his duty to the American people overcame his personal feelings.

We see additional evidence of lack of remorse or guilt in other domains of Guiteau's life. He was unconcerned about how his lifestyle impacted his father, siblings, business partners, or his former wife. In 1874, Guiteau sued the *New York Herald* for $100,000 libel for reporting on his swindling of hotels and boarding houses (*NYT* archives, July 3, 1881). Guiteau argued that the press damaged his reputation and blamed them for disclosing his dealings inappropriately. (The case went nowhere.) Perhaps one of the clearest, and yet saddest, pieces of evidence of Guiteau's emotional depravity comes from his father's private writings. Prior to the assassination, his father wrote:

> *I have been ready to believe him capable of almost any folly, stupidity, or rascality. The only possible excuse I can render for him*

is that he is insane. Indeed, if I was called as a witness upon the stand I am inclined to think I should testify he is absolutely insane and is hardly responsible for his acts. My own impression is that unless something shall stop him in his folly and mad career he will become hopelessly insane and a fit subject for the lunatic asylum. Before I finally gave him up I had exhausted all my powers of reason and persuasion, as well as other resources in endeavoring to control his actions and thoughts but without avail. I found he was deceitful, and could not be depended upon in anything, stubborn, willful, conceited, and at all times outrageously wicked, apparently possessed with the devil. I saw him once or twice when it seemed to me he was willing to do almost any wicked thing he should happen to take a fancy to.... His insanity is of such a character that he is as likely to become a sly cunning desperado as anything.... I made up my mind long ago never to give him another dollar in money until I should be convinced he was thoroughly humbled and radically changed. I am sometimes afraid he would steal, rob, or do anything before his egotism and self conceit shall be knocked out of him, and perhaps, even all that will not do it. So you see, I regard his case as hopeless, or nearly so.... (Luther Guiteau on his son, Charles, March 30, 1873)[5]

Guiteau never apologized for or indicated he felt any guilt over the assassination of President Garfield. Even on the way to the hangman, he never seemed to come close to appreciating the gravity of his crime. He perplexed an entire nation.

John Wilkes Booth, on the other hand, begged his family in his diary to forgive him for the crime he committed, and he confessed that he hated to kill but that he felt forced to act. And apparently, just before his death in the weeks following the assassination, he came to realize the fault of his act. History has recorded Booth's scheme to be one of the most vile and cowardly crimes ever committed, but he seemed to have the ability to experience remorse and guilt.

Booth 0
Guiteau 2

7. Shallow Affect

Here we are concerned with the depth, quality, and stability of an individual's emotional life. Cleckley described the psychopath as showing absolute indifference to hardships, whether they are financial, social, emotional, physical, or other, which he readily brings upon others, and even those whom he professes to love.[4] In interviews, the psychopath seldom claims he has no idea what love is. He often equates love with sex. It's not uncommon for him (the overwhelming majority of psychopaths are men) to admit that he has never really felt any emotions for anyone, other than the physical pleasures associated with sex. The careful observer will note that psychopaths will readily express feelings, emotions, and affect, but the feelings and emotion are rather limited in strength and depth of feeling. The psychopath 'knows the words but not the music'.[5]

Cleckley reported that psychopaths never experience grief, honesty, deep joy, or genuine despair. From my own experience, I would add to Cleckley's observations that the psychopath never ruminates on anything. Rumination is a process that often contributes to depression and in extreme forms to obsessive-compulsive disorder. The process of rumination is often associated with some anxiety or subjective feeling of concern or worry, and this can help precipitate change in the individual in order to reduce the anxiety. The psychopath experiences none of this. Indeed, if you ask a psychopath if he has ever worried about whether he left the house with the stove on (a common problem among those with obsessive-compulsive disorder), he will look at you like you are an alien, in stunned disbelief. Obsessive compulsiveness is completely foreign to the psychopath's way of thinking. Psychopaths are on the opposite end of the spectrum.

Certainly, with respect to his incarceration, Guiteau showed no ill effects. He actually scouted the jail where he was to be incarcerated prior to the assassination to preview the accommodations he was going to enjoy. Guards and prison food workers documented that he showed no evidence of sadness, depression, or concern during his period of incarceration prior to and up to the day of his execution. Indeed, his lack of emotional consternation regarding his execution is noteworthy. He showed absolutely no anxiety or fear of punishment. This is another hallmark characteristic of psychopaths. Punishment, real or threatened, does not alter the psychopath's behaviour. For the rest of us, a sharp tongue lashing or physical punishment will suffice to make us (re)consider our choices, and usually alter our behaviour in order to avoid such retribution in the future. For the psychopath, such punishments do not take. It's hard for people to understand this point. For the vast majority of us, real or threatened punishment is sufficient to keep us on the straight and narrow. But psychopaths seem unable to learn from punishment, severe or not. It will not change their behaviour. Take the death penalty, for example. Proponents argue it is a deterrent. But the thought of life without the possibility of parole, confined to a six-foot by nine-foot cell with very disagreeable people surrounding us, is sufficient deterrent for 99 percent of us. However, for that 1 percent who are psychopaths, even the death penalty never enters their awareness as a potential punishment for murder – it has absolutely no deterrent effect. Such was the case for Guiteau.

There is plenty of evidence that suggests Guiteau was unable to develop strong emotional attachments to people. He had few acquaintances and he was 'not able to make any friends'.[6]

Booth, in rather stark contrast to Guiteau, showed significant signs of anxiety in his life, especially during early stage performances. He maintained very close relationships with his sister and brothers prior to the Civil War. Even during the war, Booth maintained a close relationship with his sister despite the fact the Booth family was split along ideological lines. The Booths were raised in Maryland, a border state in the War Between the States. His brothers were supporters of the North while John Wilkes supported the South. Edwin, John Wilkes's older brother, distanced himself from John Wilkes because of this, and he initially disowned him following the assassination of President Lincoln. However, later Edwin lobbied and obtained rights to bury John Wilkes's body in the family burial plot so he could rest in peace with the rest of his family.

Booth attended funerals of family and close friends, and he apparently showed a normal range and depth of emotion at these and other events. He cared for and left his money and property to his siblings and their children, understanding the fateful path he had chosen.

Booth was reported to be excitable and his acting reflected that. He was known as a highly enthusiastic actor who threw himself into the characters. Many of the plays he was in portrayed the theme of killing or overthrowing an unjust ruler, as in the story of Scottish freedom fighter William Wallace. Booth's favourite role was Brutus – the slayer of a tyrant. Historians have speculated that his experiences with the plays he acted in may have contributed to his political ideology, and ultimately to his views that Abraham Lincoln was a tyrant who needed to be removed. It is worth noting that Booth was not alone in his (misguided) ideology. In any event, there is considerable evidence that Booth was able to experience a normal depth and range of emotion.

> Booth 0
> Guiteau 2

8. Callous/Lack of Empathy

Guiteau was known through the city of Chicago as a 'pettifogger' – a vicious, wild character. He'd been barred from working in the courts, which precipitated his departure from Chicago and move to the East Coast. He was described by his father, brother, brother-in-law, sister, former wife, friends, and business associates as having a wicked temperament. In his swindlings, he was impervious to sentiment or feeling for the victims. His former wife reported that she never once, in her five years with him, heard him say that he was sorry for his behaviour; he apparently did not suffer pangs of regret on account of his dishonesty.[7] His natural disposition was reported as 'ugly'. He delighted in controlling others. He mentally and physically abused his wife daily and demonstrated callousness and a lack of empathy to individuals throughout his life.

One of Booth's letters helps to better understand his empathy and understanding. Is this the letter of a callous man?

TO WHOM IT MAY CONCERN:

Right or wrong, God judge me, not man. For be my motive good or bad, of one thing I am sure, the lasting condemnation of the North. I love peace more than life. Have loved the Union beyond expression. For four years have I waited, hoped and prayed for the dark clouds to break, and for a restoration of our former sunshine. To wait longer would be a crime. All hope for peace is dead. My prayers have proved as idle as my hopes. God's will be done.... I know how foolish I shall be deemed for undertaking such a step as this, where, on the one side, I have many friends, and everything to make me happy, where my profession alone has gained me an income of more than twenty thousand dollars a year, and where my great personal ambition in my profession has such a great field for labor. On the other hand, the South have never bestowed upon me one kind word; a place now where I have no friends, except beneath the sod; a place where I must either become a private soldier or a beggar. To give up all of the former for the latter, besides my mother and sisters whom I love so dearly, (although they so widely differ with me in opinion,) seems insane; but God is my judge. I love justice more than I do a country that disowns it; more than fame and wealth; more (Heaven pardon me if wrong,) more than a happy home.... My love (as things stand to-day) is for the South alone. Nor do I deem it a dishonor in attempting to make for her a prisoner of this man, to whom she owes so much of misery. If success attends me, I go penniless to her side. They say she has found that 'last ditch' which the North have so long derided, and been endeavoring to force her in, forgetting they are our brothers, and that it's impolitic to goad an enemy to madness. Should I reach her in safety and find it true, I will proudly beg permission to triumph or die in that same 'ditch' by her side. (A Confederate doing duty upon his own responsibility. J. WILKES BOOTH. REPORT FOR THE YEAR 1864)

Booth had adopted the racist ideology of the time. Personally, I find it very difficult not to score a racist who endorses slavery high on *Callous/Lack of Empathy*. However, we have to review the rest of Booth's life. In it we find little evidence of any callousness or inability to empathize with others. Since we find evidence in only one domain of his life, we must score him in the low to moderate range.

Booth 1
Guiteau 2

9. Parasitic Lifestyle

Guiteau frequently borrowed money from friends and relatives, lived with them for long periods of time, often made no effort to obtain a job, and apparently never paid anyone back. He borrowed money from others to attend university well past when he dropped out.

When I lecture to undergraduate audiences, I note that most of us, on at least a few occasions, have asked our parents for a little extra money. But if you go to your mother and father repeatedly, to the point where they are having problems paying their bills, you begin to fulfil the criteria for this trait.

Guiteau continually relied on others. He was described by his father as 'lazy beyond degree'. He pretended to be a preacher so that he could get free travel on trains. Guiteau epitomized a parasitic orientation in all relationships and business dealings.

Booth, as far as I have found, never borrowed money from anyone. He donated to charities. There is extensive documentation regarding his reliance on hotels while on the road, but there is no evidence suggesting that he failed to pay his bills. He left bonds, cash, and land to his sister and her family in his last will and testament.

Booth 0
Guiteau 2

10. Poor Behavioural Controls

Guiteau once threatened to kill his sister with an axe (although he later claimed he was only horsing around). He reportedly drew a pistol on a man in Rochester, New York. At age eighteen, he got in a fight with his father. His father described him as someone with a wicked, explosive temper. Stories of physical altercations on the part of Guiteau are relatively rare, but verbal and aggressive outbursts are common. Others described him as intensely high-tempered, becoming angry upon the most trifling provocation, or no provocation at all. He often lost control of himself. Perhaps one of the most salient and documented examples of Guiteau's poor behavioural controls comes from his behaviour at trial. He verbally interjected during all phases of the trial, during jury instruction, testimony, routine lawyer interactions, and so on. He would frequently pound his fists upon the table, raise his voice, even stand up and shout, often to the chagrin of the prosecutor. The judge was notably lenient in this regard. However, the shouting, screaming, or fist pounding wasn't done in any delusional or psychotic sense. It was just poorly regulated behaviour. The general public was shocked by his antics, but Guiteau's wife was unfazed, stating that he was always like that.

Booth has been described as rather impulsive, often acting without planning or thinking. Some accounts discuss Booth as an excessive drinker and include stories of uninhibited behaviour. But it does not appear that Booth was hotheaded to the degree that Guiteau was.

> Booth 1
> Guiteau 2

11. Promiscuous Sexual Behaviour

Guiteau's wife, Annie Bunn, successfully won a divorce from Guiteau by presenting Guiteau's mistress, Clara Jennings, at the proceedings. On the stand Jennings admitted to having an affair for years with Guiteau.

In the language of the times, Guiteau was variously described as a man of many vices, and a man who supported 'vile' women. In one report he frequented brothels, which resulted in his contracting venereal disease. It's difficult to document Guiteau's sexual proclivities over the course of his life, but there appears to be enough evidence to safely warrant a high score.

Booth scores high here too. His sexual liaisons were legendary – he rarely went home alone after an acting performance. But his attitude towards sex was not dissimilar to his brothers' or father's, or to the attitude of many other men of the time. Nevertheless, Booth had a very cavalier, if straightforward, attitude about sexual relations. He simply chose to engage in as many liaisons as possible, which merits a high score on this trait.

> Booth 2
>
> Guiteau 2

12. Early Behavioural Problems

Charles Guiteau was described by a family friend as a fear and terror to his parents growing up. He possessed a mean and stubborn disposition. His father could not control him and (correctly) predicted Charles would bring disgrace to the family. He apparently tried everything he could to make a good man of Charles, but to no avail. By most accounts the family was considered highly respectable, and Charles's five siblings were apparently raised without incident. Charles was the black sheep of the family.

His father was considered a model parent by most. Alternative accounts suggest that his father was authoritarian and highly religious, even to a pathological degree. It is not clear how Guiteau's father disciplined Charles. But Charles was known throughout the neighbourhood as a 'miserable rascal'. A family physician had diagnosed Charles as an imbecile, echoing the term *moral imbecility*, popular at the time (a euphemism for *psychopathy*).

Perhaps the most salient historical evidence about his early years comes from a confession Charles provided the Oneida religious community prior to being accepted into their commune in upstate New York at age eighteen. He confessed that he had, from very early years, been a very disobedient boy to his father; that he had, while a clerk in a business house, between his thirteenth and seventeenth years, robbed his employer's money drawer repeatedly of considerable sums of money; that he had frequented brothels; that he had contracted venereal disease; and that he had been addicted to self-abuse to the extent of seriously injuring his health. Charles reported a serious fight with his father around age eighteen, to the point where he wanted to hurt his father.[8]

With this symptom, we are generally concerned with very early childhood, and I have been able to find only limited evidence of Guiteau's early years. The little evidence we do have suggests that Guiteau demonstrated problem behaviour at an early age. But without being able to interview Charles about his early childhood and in the absence of other solid sources of information, such as school records, social worker reports, and accounts from older siblings, it is best for us to omit this item. The practical consequence of omitting an item is that the total score on the Psychopathy Checklist is prorated, based on the rest of the item scores.

Today, a high score on this item is typically reserved for children who are taken from the home by social services before the age of twelve, even when the parents are taking care of other children without serious incident. The child's behaviour should be relatively independent of the environment.

With respect to Booth, there is much documentation about his early years from his siblings, and there is very little evidence he had anything other than a typical childhood. Edwin reported that John Wilkes was the parents' favourite (of ten) in the home (only six would live to adulthood – a cholera epidemic killed three siblings in a single month in 1833 and another died from smallpox four years later). His sister recounted a story of fireworks gone wrong, in which a passerby was slightly injured. When the constable arrived, John Wilkes confessed to setting off the fireworks, but he refused to name his friends and brothers who contributed, solely taking the blame himself. He practised and maintained an interest in music, musicals, and theatre from a young age. He shows no evidence of aberrant behaviour early in life.

Booth 0
Guiteau omit, prorate final score

13. Lack of Realistic, Long-Term Goals

Here we are concerned with whether the individual can make realistic plans for the future, stick to goals, and carry them out. Charles failed on all attempts. His plans for the future were always misguided and often grandiose (see Item 2). At various times Guiteau had applied to be the minister to Austria, consul to Liverpool, and consul-general to Paris, yet he had no real hope of ever obtaining such posts.

Guiteau planned to (re)marry rich, telling his former wife he had numerous wealthy women at his beck and call and, once married, would be in a position to pay her the alimony to which she was so deservedly entitled. Several years prior to the assassination, he wrote the following letter to then–General Garfield:

> *Dear General:*
>   *I, Charles Guiteau, hereby make application for the Austrian Mission. Being about to marry a wealthy and accomplished heiress of this city, we think that together we might represent this Nation with dignity and grace. On the principle of first come first served, I have faith that you will give this application favorable consideration. CHARLES GUITEAU.*

Apparently, Garfield was shown the letter, and several times afterward referred to it humorously as an illustration of unparalleled audacity and impudence.

Guiteau had an unbridled, unabashed, and unwavering penchant for making unrealistic long-term plans.

During the Civil War, John Wilkes Booth retired from performing in the theatre and began investing in land and oil. Apparently, these investments did not do well, and he had made plans to return to the theatre to continue to support himself.

Booth and his conspirators had initially plotted to kidnap President Lincoln to force a prisoner swap for Southern troops being held in the North. As the Confederacy started to collapse, the plan quickly escalated to a coup d'état. The plan that fateful night was to assassinate President Abraham Lincoln, Vice President Andrew Johnson, and Secretary of State William Seward. The conspirators' hope was that the US government would go spinning out of control and the South could regain traction or force a revised settlement. Booth successfully executed Lincoln; his coconspirators were not as adept. One coconspirator gruesomely stabbed Seward in the head several times, but Seward miraculously survived, and the assassin assigned to Johnson lost his nerve. Thus, although the plan had goals, it is not clear what would have happened if their ploy had succeeded. It's actually quite scary to think about what might have happened. In any event, Booth had no problems making plans in the various domains of his life, and he was effective at carrying them out.

Booth 0
Guiteau 2

14. Impulsivity

Impulsivity is a multifaceted construct. In assessing this trait we are mostly concerned with behaviours that are unpremeditated, unplanned, or opportunistic.

Despite normal or perhaps even above-average intelligence, Guiteau was never able to maintain a steady job, despite the 'ability' to do so. He moved from job to job, rooming house to rooming house, never thinking ahead. In relationships, he did not maintain either close personal friends or romantic relationships. He moved frequently, and at various points in his life lived in California, Maine, Illinois, and New York.

Guiteau was clearly capable of designing and completing con jobs and swindles. He also planned and stalked the president, going so far as to practise shooting his new handgun prior to the assassination. However, it is clear that these plans were always short-lived and not well manufactured (he never really made a lot of money, and he got caught right after shooting Garfield). Given that, Guiteau scores high on this item.

Booth described himself as impulsive in thought, but there is little evidence that he acted without planning or would be considered impulsive. However, there are enough stories regarding his drinking behaviour, the occasional barroom fight, and a few other examples that merit a moderate score on this item for Booth.

> Booth 1
> Guiteau 2

15. Irresponsibility

This item is generally concerned with a person's sense of responsibility in all domains of life. Guiteau had no loyalty with respect to any area of his life, in relationships with family or in romance, education, or business dealings. According to newspaper stories, Guiteau owed money to nearly every hotel in New York City, Boston, and Chicago for skipping out on room, beverage, and food charges. He was charged with paying alimony, but his ex-wife reported he never paid a dime. He was disbarred in Chicago for fraud over a settlement agreement for some $275. He borrowed money from everyone and anyone who would lend it to him on his 'word of honor', but there are no accounts of him ever paying back anyone. He is the epitome of irresponsibility.

Booth was motivated singularly by his sense of duty to the Confederacy. It was this sensibility that pushed him into the violent act he committed. It is impossible to justify Booth's act, but it is clear his motivations were driven out of a perverted sense of loyalty and duty to the South. In the rest of his life, we see no evidence of irresponsibility.

> Booth 0
> Guiteau 2

16. Failure to Accept Responsibility for Own Actions

Guiteau's most famous quote is, 'The doctors killed Garfield; I just shot him.' In other words, he failed to take responsibility for killing the president. Guiteau blamed his wife for their stark financial situation, saying he should have married a rich woman. He blamed the *New York Herald* for destroying his reputation by publishing facts about his questionable business dealings. When he started a newspaper called *The Daily Theocrat* with an investment from the Oneida commune and the business venture failed, he blamed the Oneida commune for not giving him enough money. He sued them, unsuccessfully, and resorted to blackmail, threatening to sell stories about the inner workings of the community to newspapers. Guiteau's ex-wife claimed that he was unable to accept responsibility for anything he ever did.

Booth, in the letter reprinted earlier, owns his crime. He accepts responsibility, even his own death, for what he has done. He was willing to, and did, sacrifice himself for his beliefs, however misguided history might record those beliefs to be. This is the essence of accepting responsibility for one's behaviour.

Booth 0
Guiteau 2

17. Many Short-Term Marital Relationships

The flavour and intent of this item is to address the fact that the psychopath is willing to jump into (often without thinking) marriage and then jump out (again often without thinking). Some psychopaths I have interviewed have been married six or seven times.

Guiteau has only one marriage that we know of, to Annie Bunn. Perhaps his cons and swindles left no time for Guiteau to engage in other significant romantic relationships.

Booth never married, perhaps because he died at a relatively young age, twenty-seven. Booth was reported to be engaged once, but that does not warrant a score on this item.

>   Booth 0
>   Guiteau 0

18. Juvenile Delinquency

The symptoms of psychopathy start early in life and accelerate in adolescence. This item is concerned with the severity of delinquency during the teenage years. Many adolescents engage in moderate levels of troublemaking, but the vast majority of teens mature out of those behaviours without serious incident. Here we are looking for evidence of severe and chronic antisocial behaviour that would likely lead to arrest and serious charges as a youth.

With respect to Guiteau, one of the pieces of information we have is his confession for robbery, use of prostitutes, and the occasional assault during his teen years. Reports from family, neighbours, and friends all suggest Guiteau was unmanageable and a terrorizer of the town in which he grew up. Many of these stories were documented prior to the assassination, minimizing the concern that post-assassination reports would inflate the antisocial behaviour of his youth. It is safe to score him high on this trait.

I was unable to find any reports indicating that Booth had any serious trouble with the law as a teenager. All collateral reports from siblings and friends reported no significant antisocial behaviour during adolescence.

>   Booth 0
>   Guiteau 2

19. Revocation of Conditional Release

Here we are looking for repeated failure when given an opportunity to redeem oneself. Most criminal justice systems have mechanisms to attempt to curb antisocial behaviour. Probation, in lieu of jail time, is one such mechanism. The point of probation is a warning shot across the bow – don't get in trouble again or you are looking at a stiff penalty. The first-time offender is hopefully taught a lesson and learns from his or her experience and from the threat of punishment. As discussed, psychopaths fail to learn from experience, and the threat of punishment has no bearing on their future behaviour. This item helps capture that trait. It also assesses the willingness of an individual to breach trust from a source other than family and friends. When a government releases someone early from prison, in the case of parole, some readily fail that trust.

Guiteau has both breach of bail and failure to appear charges. Multiple incidents of breaching warrant a high score.

For Booth we have no evidence he was ever arrested, so we have to omit this item and prorate his final score.

> Booth omit, prorate total score
> Guiteau 2

20. Criminal Versatility

Guiteau's list of transgressions is a long one: murder, fraud, theft, assault, robbery, threatening use of handgun, possession of a weapon, forgery, jumping bail, assault on a law enforcement officer (fight with a guard while in jail), unlawful confinement (reports during his divorce described at least one event when he forced his wife into a closet and held her there until she nearly suffocated before finally letting her out), vagrancy, and public drunkenness. Guiteau is an equal opportunist when it comes to antisocial activity.

This item chronicles not just how much antisocial activity has been committed, but also a willingness to engage in a wide variety of criminal activity. The psychopath tends to engage in many different types of criminal activity, which differentiates the psychopath from the typical offender who just commits, say, burglaries to support his or her drug addiction. It's the variety here that is important.

Booth had a few vices, and while he did commit murder in

assassinating Lincoln, there is no evidence of criminal activity in multiple domains of his life.

> Booth 0
> Guiteau 2

Summary

This exercise illustrates how scientists assess psychopathic traits, through the lens of two nineteeth-century presidential assassins. Both men committed heinous crimes with which their names will for ever be associated. However, as this analysis has revealed, the two men could not be more different with respect to their psychopathy scores.

Guiteau scores in the 99th percentile of psychopathy (37.5 out of 40) (noting the half point is due to the prorated item). In other words, of the thousands of individuals studied with the Psychopathy Checklist, Guiteau is an exemplar for nearly every trait – close to a perfect score. In addition to meeting nearly all the modern criteria for psychopathy, Guiteau meets all of Cleckley's criteria as well. Notably, Guiteau never suffered from delusions or hallucinations or experienced anxiety – including on his way to the gallows. Even his assassination of President Garfield was poorly motivated. Dr Spitzka, who testified for the defence during the assassination trial, went a little beyond moral insanity and stated that Guiteau was 'a moral monstrocity'. Guiteau's behaviour at trial was theatrical, to say the least. He testified using poems he had written, he openly and repeatedly sought to represent himself in court, and he berated his defence team. Sentenced to death, he requested an orchestra play during his execution (the judge denied the request). On his way to the hangman, Guiteau continued to display psychopathic exuberance, as he danced up the stairs, waved to the audience, shook hands with the executioner, and, as a last request, recited a poem he had written.

Guiteau's lack of motive left an entire country seeking answers.

Following his conviction, Guiteau was hanged in effigy by the citizens of Flint, Michigan, where he grew up and where people had had to deal with him for years – a poignant end to a tragic story.

Booth, on the other hand, scores only 8.4 out of 40, a low score. While Booth's psychopathy score is twice as high as the average male (the average man will score a 4 out of 40 on the Psychopathy Checklist), he is well below average for a criminal. Prior to his crime, he is the sort of person I might have had a beer with, with the hope I could talk him out of his misguided political ideology. Nevertheless, Booth has gone down in history as the man who committed one of the most notorious acts in American history and robbed the country of one of its greatest presidents. For this, he will remain infamous. But he cannot be labelled a psychopath in the clinical sense. Instead, we are left with calling him what he was – an assassin.

chapter 4

The Psychopath Electrified

ғᴀᴄᴛ: seventy-seven percent of psychopaths in the united states are incarcerated.[1]

Nineteen ninety-eight. My Sunday morning began with a sixty-minute commute through the rain to the home of the maximum-security treatment programme for Canada's most notorious violent offenders. This was a special day as a new cohort of inmates was being transferred in to start treatment. I was excited at the chance to interview twenty-five new inmates and get them signed up for my PhD research studies.

At that point I had been working in Canadian prisons for more than five years. I had interviewed hundreds of inmates, many of them psychopathic. A few had even achieved perfect scores on the Psychopathy Checklist.

On prison workdays I always arrived early, before the inmates were required to stand in their cells for the morning count at 7 a.m. I worked through lunch and stayed until the guards kicked me out at 7 p.m., when the inmates were locked down for the night. I was in heaven. I was living my dream, interviewing psychopaths on a daily basis. It was absolutely captivating. At the same time, I worried that in a couple years I was going to complete graduate school and leave this place. I was concerned that I would never be able to work at such an amazing facility again, surrounded by such supportive staff and, frankly, such co-operative inmates.

So I worked a lot. I worked weekends because the inmates have more free time then than during the week. I skipped the winter term break and spent Christmas in prison. Some of my friends said that I was a workaholic and I needed to learn how to take a break. But how could I stop? Especially when there was a whole new crop of inmates arriving. My PhD supervisor, Dr Robert Hare, told me that he feared that I might be manic. He was somewhat calmer after he discovered that this was not a transient episode. I had worked this schedule for years now.

I had a fresh stack of printed Psychopathy Checklist interviews; the questions and their timing and delivery had been perfected through hundreds of interviews. I set up the video camera on its tripod, unwrapped a fresh videocassette, and loaded it into the bay on the side of the camera. I headed up to the inmates' housing units, twirling my brass key in my hand. I caught myself whistling a tune. I was excited to meet the new inmates. However, this Sunday would exceed my wildest expectations. Indeed, it was a day I would never forget.

I arrived at the housing unit before the inmates had left their cells. I entered the nurses' station and fired up the coffeemaker.

The inmates' cells opened and they rushed for the showers or the TV room. It was American football season and the East Coast games were just starting. The inmates crowded into the TV room. I leaned against the door frame, watching the TV to see if I could catch a glimpse of the latest highlights. I flashed back to my own football days, then I realized that I was standing in the way of a violent offender who wanted to grab the last seat in the TV room. He gently nudged me aside and took his seat.

And then suddenly there was tension in the air. I felt it on the back of my neck before I was even conscious of what was happening. The inmates milling around had slowed, the sound of their feet hitting the cold concrete floor halted, the TV seemed to get louder, and all of a sudden I was acutely aware of the steam from the hot coffee in my mug spiralling up towards my nose.

An inmate had exited his cell completely naked and started walking up the tier. I noticed him out of the corner of my eye. He passed the TV room, shower stalls, and empty nurses' station and proceeded down the stairs to the doors that led to the outside exercise area. Some of the inmates turned slightly after he had walked by to take a look at him. Others tried not to move or look, but I could see they noticed. The inmates were as confused as they were anxious. What was he doing?

The naked inmate proceeded outside into the rain and walked the perimeter of the short circular track. He walked around the oval track twice. The TV room was on the first floor and the inmates had a good view of the track. Some of the inmates peered outside and watched him. Everyone was distracted; no one spoke. We were all in shock.

The inmate returned, still naked, and walked up the stairs to the second-floor tier and then down to his cell. The tension around the TV room grew. The inmate quickly emerged from his cell with a towel and proceeded to the showers. He walked down the middle of the tier as inmates slowly moved out of his way or retreated into their cells. Other inmates appeared to talk to one another, but they were clearly trying to avoid any direct eye contact with him. I noticed one of the biggest inmates had subtly slowed his pace so that he would not cross the path of the new inmate.

The naked inmate took a quick shower and returned to his cell; there was a slight swagger to his stride. He was not particularly big, but his physique was ripped.

I had to interview him. I took a gulp of coffee and then walked towards his cell.

The first name written on masking tape above his door was 'Richard'.

'Good morning. I'm the research guy from UBC. We are conducting interviews and brain wave testing on the inmates in treatment here. Would you be interested in hearing more about it?' I asked.

'Sure,' came the reply out of the dark cell.

'All right, then. Why don't you get dressed and grab a bite to eat, and I'll come get you in about thirty minutes. We'll do the interview downstairs in my office.'

I returned to the nurses' station and had a couple more cups of coffee. I wanted to make sure I was fully awake when I interviewed Richard.

Richard had dressed in classic prison garb: blue jeans, white T-shirt, and dark green jacket. He sauntered down the stairs and through the covered outdoor walkway to the mess hall for breakfast. He returned to his cell after about fifteen minutes. I couldn't wait; I went down early to get him.

He followed me to my office and he plopped down in the chair opposite from me.

Before I could get the consent form out of the drawer, he stared at me and said: 'You ever need to push that red button?' He was referring to the big red button in the middle of the wall; when depressed, it signalled distress. A buzzer would go off in the guard bubble down the corridor.

We were both about the same distance away from the button. I realized that I might not be able to reach the button before he could get to me. My mind quickly turned to figuring out a new way to organize the office so that I was closer to the button than the inmates being interviewed.

'No,' I replied. 'In the five years I've worked here, I've never had to push the button.' I threw the *five years* in to let him know that I had some experience behind me. I didn't want another con game played on me by inmates who thought I was a 'freshman' in maximum security.

Without saying another word, he leaped up and slammed his hand on the button. I didn't have time to react. He returned to his seat as quickly as he had jumped up.

'Let's see what happens,' he said calmly, leaning back into his chair.

Over a minute later, we heard doors being slammed open in the distance and the unmistakable sound of running footsteps.

I had thought about getting up and opening the door for the guards, but I would have had to pass by Richard to get to the door. So I just sat in my chair and waited. Richard looked around calmly at the computers, files, and books that had accumulated in my office over the last five years. The office I had commandeered had a little side closet, about five by five feet, where I had put all my brain wave recording equipment. I had set up the rest of my office so that I could monitor the computers while the inmates sat in the little closet on a comfortable chair and submitted to my EEG studies. The downside to my office configuration was that during interviews the inmate was closest to the exit. Had I become complacent? Should I have rearranged my office and stuck to the rule Dr Brink mentioned to me on my first day – to always put my seat closer to the door in case I pissed off one of the inmates? I hoped I hadn't made a fatal mistake.

These thoughts raced through my mind during the eternity it took the guards to get from their station down the stairs and to the end of the corridor, where my office resided. It was a Sunday morning and I was one of the only staff people at the facility. The guards' response time felt glacial.

A key was jammed into my door and then it was flung open; two guards entered, panting and out of breath, and stared at us.

Richard turned calmly in his chair and said to the guards: 'What's the problem?'

'Someone pushed the alarm button,' the guard stammered. 'Everything okay?' His question was directed at me.

'Oh, I must have accidentally pushed it when I took my coat off,' Richard answered. 'Everything is just fine; we are just doing the research interview here.'

'Okay,' the guard said. 'Don't do that again.'

I just nodded. I was having trouble speaking.

The guards pulled the door closed and Richard turned and looked at me.

'They call me *Shock Richie*,' he said. 'And I'm going to shock you too.'

Mustering as much inner strength as I could, I replied: 'I'm looking forward to it; I'm here to be shocked. Take your best shot.'

Shock Richie smiled.

Prison is never boring, I thought.

We completed the consent form and then I started the Psychopathy Checklist interview with a question I would never ask any other inmate in my career.

'Why did you walk naked out in the rain?'

'Well, I arrived last night. You have to make an impression on the other inmates right away when you get shipped to a new place. I saw you standing there by the TV room. You noticed how all the other inmates got a bit nervous when I walked by. Even the big ones get nervous when you do shit like that. You just got to establish yourself right away. If you don't, then inmates think they can test you.' He stared quite matter-of-factly at me; the emptiness in his eyes was unnerving.

'When I do stuff like that, inmates don't know what to think. I'm unpredictable. Sometimes I don't even know why I do what I do. I *just do it.*'

My mind was racing again. I completely agreed with his logic, albeit twisted; he had already established his dominance at this prison. He was going to score high on at least a few Psychopathy Checklist items. Nike probably never envisioned a psychopathic inmate embracing their slogan *Just Do It* in a manner quite like this.

'You've been working here for five years?'

'Yes, since I started graduate school,' I replied.

'Interviewed lots of guys, right?'

'Yes, hundreds of them.'

'Well, you ain't never met anyone like me,' he said.

'Really? What makes you so special?'

'I've done shit you can't even imagine. I'm gonna shock you like I shock everyone,' he stated calmly. 'Let's get on with it.'

Richie enjoyed doing bad things. He was only in his late twenties when I interviewed him, but he had a rap sheet like no one I had ever interviewed before. As a teenager he had committed burglary, armed robbery of banks and shops, arson for hire, and all kinds of drug-related crimes from distribution to forcing others to mule drugs for him. He would force women to hide plastic baggies of cocaine in their body cavities and transport them across borders and state lines and on plane flights. One of Richie's girls got a baggie stuck in her vagina. Richie used a knife to 'open her up a bit' so he could retrieve his drugs. He said he didn't use her again after that. When I asked him what he meant by that, he said that he didn't use her for sex; she was too loose now, and she lost her nerve about carrying drugs.

Richie smiled as he told me a story of a prostitute he had killed for pissing him off. He actually seemed proud when he described wrapping her up in the same blanket he had suffocated her with so he could keep all the forensic evidence in one place. He put her in the boot of his car and drove out to a deserted stretch of road bordered by a deep forest. Chuckling, he told me he was pulled over by a highway trooper because he was driving erratically as he searched for a dirt road to drive up so he could bury the body in the woods.

'So the cop pulls me over and comes up to the window and asks me if I have been drinking alcohol. I lied and said no. I told him that I just had to take a piss and I was looking for a place to go. But the cop gave me a field sobriety test anyways. I figured that if I didn't pass the test, I would have to kill that cop. Otherwise, he might open the trunk and discover the body. The cop didn't search me when I got out of the car, and I was carrying a knife and a handgun. I'm surprised that I passed that field test since I had had a few drinks that night. I was planning to beat the cop senseless and then I was going to put the girl's body in the backseat of the cop's car. Then I would shoot him in the head with his own gun and make it look like a suicide after he accidentally killed the prostitute while raping her in the backseat of his cruiser. Everyone would think it was just another sick dude.'

The irony of his latter statement was completely lost on Shock Richie.

The cop proceeded to point out a dirt road just up the way where Richie could pull over and take a piss. It was fascinating that Richie could remain calm enough not to set off any alarm bells for the cop that something was amiss. After all, Richie had a body decomposing in the boot of the car. Yet apparently, Richie showed no anxiety in front of the cop. Most psychopaths like Richie lack anxiety and apprehension associated with punishment.

Richie turned up the dirt road the cop pointed out to him and drove in a ways. He pulled over, parked, and removed the body from the car.

'I had all these great plans to carry the body miles into the woods and bury it really deep so nobody would ever find it. But it's fucking hard to carry a body. You ever tried to carry a body?' he asked.

'No, I don't have any experience carrying dead bodies,' I told him.

'Well, it's a lot of work, let me tell you. So I only got about a hundred yards off the road and just into the trees before I was exhausted. Then I went back and got the shovel from the car. I started digging a huge hole.'

He looked up at me with those empty eyes and asked: 'You know how hard it is to dig a hole big enough to bury a body?'

'No,' I answered, 'I don't have any experience digging holes to bury bodies.'

'Well, it's harder than you might think.' He continued, 'So I took a break from digging and noticed that my girl had rolled out of the blanket and her ass was sticking up a bit. So I went over and fucked her.'

He got me. And he knew it.

'Surprised ya with that one, didn't I? Told ya.' He was proud of himself.

As my stomach turned, I managed to utter a reply: 'Yes, you got me with that one.'

'She was still warm, ya know, and I just got horny. What's a guy gonna do? She was always a nice piece of ass.

'So I finished up there a bit and thought maybe I should burn the body and the evidence, but in the end I decided to just cover her in dirt.

'Ah.' He starts laughing. 'I had all these great plans to carry her miles into the woods and dig this monster hole so nobody would ever find her.'

A couple weeks later, a couple of hikers discovered the body. Shock Richie read about it in the newspapers, but he was never charged with the murder.

I thought back to my first day in prison, reading the inmates' crimes written above their history binders in the nurses' station. So there *was* a difference between rape/murder and murder/rape. Part of me still wished I didn't know.

Richie admitted that he had no need for friends. He'd really never been close to anyone in his life. He preferred to do everything on his own. He also didn't trust anyone. I believed him. Richie had no friends in prison, he had no visitors, and all the other inmates said he could not be trusted and he knew not to trust them in return.

He had lived a life supported by crime, never had any vocational training, and never made even a passing attempt at any other lifestyle. He made most of his big scores by taking down rival drug pushers. He would set up deals in different towns and then rob and sometimes kill the other person. Richie had no fear or hesitation with killing. Richie also had more than a dozen fake names and accompanying identification.

For a long time he was a pimp. He used to corral runaways into working for him. He would get them hooked on drugs and then make them work the streets. He'd killed more than a few prostitutes. He saw people as objects, things to be manipulated; we were there just for his entertainment.

When Richie had been released the last time from prison, he was taken in by his older brother. His older brother was not a criminal. He was on the straight and narrow. After a few months of Richie bringing home prostitutes and doing drug deals at the house, his brother had told Richie he had to stop or he was going to kick him out. They argued, but Richie never tried to change his behaviour. Finally, his brother had had enough. He picked up the phone to call the police to have him arrested for drug possession. 'I was high,' said Richie, 'but not more than usual. I got the jump on him and beat him with the phone. While he was lying there dazed on the floor, I ran into the kitchen and grabbed a knife. I came back and stabbed him a few times.' He looked up at me intently to see if I was shocked.

'Continue,' I said.

'I figured that I would make it look like somebody had come over and killed him as part of a drug deal gone bad. Then I thought that maybe I should make it look like my brother had raped one of my girls and one of them had stabbed him.' By girls he meant the prostitutes in his 'stable'.

After killing his brother, he went out and partied for a day or two. Then he came back home with a prostitute whom he planned to stab, and then put the weapon in the hand of his dead brother. He was going to put them both in the basement and make it look like his brother died quickly during the fight and the girl died slowly from stab wounds.

While he was having sex with the prostitute in the living room, she said she smelled something funny.

'You ever smell a body after it's been decomposing for a couple days?' he asked.

'No,' I replied, 'I don't have any experience smelling decomposing bodies.'

'Well, they stink. I recommend getting rid of them fast.'

After having sex, he intended to lure the girl down into the basement. But the prostitute excused herself to use the toilet and she jumped out the window and ran away. Later that evening the police showed up at his door and asked to come inside. Apparently, the prostitute recognized that odd smell to be that of a decomposing body. She had good survival instincts.

Richie told the cops he had been away from the house partying for a few days. He didn't know that his brother had been killed. Confessing to being a pimp and drug dealer, Richie told the officers that he owed a lot of people a lot of money. He gave them a list of a dozen or so names of potential suspects.

The police eventually arrested Richie. Through his attorney, Richie received a plea deal. He pleaded guilty to manslaughter and was sentenced to seven years in prison. He'd served six and was scheduled for release when he completed the treatment programme.

Richie had a few more zingers he hit me with that day. He had indeed met my challenge. When I got home that evening, I opened a bottle of wine; it was empty before I knew it.

By the end of the following week, I had completed interviews of most of the new inmates and scheduled them for EEG studies. Shock Richie was the first on my list. I had to know what his brain waves looked like.

Mystery Brain Wave

First, a bit of background on brain wave activity. Neurons in the brain act like millions of little batteries. The electrical field that these batteries generate can be recorded from the surface of the head using sensors called *electrodes*. Brain wave activity is known as the electroencephalogram, or EEG. It is common for EEG to be used to evaluate and screen for clinical disorders, from detecting electrical seizure activity in epilepsy to identifying abnormal EEG rhythms that characterize sleep disorders.

EEG can also be recorded and analysed on a computer to yield details about how the brain processes information. One technique examines how certain stimuli, like words or pictures, are processed in the brain. Scientists present these words or pictures on a computer screen to research participants while their EEGs are recorded. Using a computer algorithm, scientists average the second or so of brain activity following the presentation of each class of pictures or words. This so-called event-related potential (ERP; pronounced 'erp', as in Wyatt Earp) describes the brain waves associated with processing types of stimuli. Scientists who publish ERPs are affectionately called *ERPers*. In graduate school, my classmates nicknamed those of us who did this work in prisons 'the ERPer mafia'.

ERPs are typically about one second long. Scientists generally break ERPs into early, middle, and late components. Early components, the first 200 milliseconds or so, typically reflect sensory processing and attentional demands. Big early components mean either the word or picture grabbed the attention of the person or the person was focusing hard to study the word or picture. Middle components happen between 200 and 500 milliseconds after the onset of a stimulus and reflect working memory, contextual updating, and motor processes associated with generating a response to the stimulus. The more memory is engaged, typically, the bigger the middle components. Finally, the late components are considered to last from 500 to 1,000 milliseconds and reflect evaluative processes in the brain. If you really ruminate on something, it often leads to a bigger late component.

The various peaks and troughs of the ERP are labelled by their ordinal position and polarity. The first negative peak is called the *N1*, the third positive peak is called the *P3*. Positive and negative deflections are equally meaningful. If the little battery (collection of neurons) is facing up, you get a positive wave, if the battery is facing down, you get a negative wave on the scalp.

I got my start using ERPs in a rather interesting way. I helped in a study using ERPs to evaluate the hearing range of killer whales. Back at UC Davis I had attended a lecture by a PhD student, Michael Szymanski, who was developing a mobile ERP system to record the brain wave activity of killer whales at the nearby Marine World Africa USA in Vallejo, California. I volunteered to help him develop the mobile whale ERP system. I had four years of experience working at UC Davis Veterinary School's Department of Anesthesia, recording and monitoring brain waves while animals were undergoing surgeries. Michael was happy to have the extra technical help, and we became lifelong friends.

We designed sensors that fit in three-inch suction cups that would stick to the killer whale's head and record ERPs. Marine World animal trainers had taught the whales to hold a position at the side of the tank so we could place the suction cups on their heads. Whales don't have a pinna or earlobe to orient to sounds. Instead, sound underwater is transduced through their jaw. So we played the sounds to the whales underwater with a special speaker, aptly called a *hydrophone*.

The study was quite a technical challenge to complete, but after two years of designing and redesigning, testing and more testing, we figured out how to record the killer whales' brain responses to sounds. Our study capitalized on the fact that the brain automatically responds to sounds presented in frequencies it can detect. This so-called auditory brainstem response (ABR) is very fast, just seven milliseconds long, and represents the firing sequence of the first seven nuclei of the auditory track in the brain as sounds are detected. So if we saw an ABR, we knew the whale could hear the frequency of the sound. Conversely, if we didn't see an ABR, then we knew the whale could not hear that sound frequency. The ABR technique is commonly used to detect developmental hearing problems in children, to detect damages in hearing following head injury, or to assess hearing loss after too many loud rock music concerts.

We found that killer whales can hear frequencies that are ten times higher than those humans can hear. Killer whales can also hear frequencies much lower than humans. This work led to changes in how the US Navy conducted ocean experiments. The navy's studies were modified so that the noise envelope produced by their experiments would not interfere with the optimal hearing frequencies of killer whales. The brain waves from Yaka and Vigga, the two killer whales at Marine World, were published in a prestigious journal.[2] Twenty years later, they are still the largest animals from which brain waves have ever been recorded. Michael and I are very proud of the work we did and the conservation efforts that resulted to protect whale hearing and communication.

My initial training in killer whale ERPs also introduced me to many of the paradigms used to elicit robust brain waves in animals and in humans. My favourite task to ERP is also one of the simplest. It's called the *Oddball Task*. In it, participants are presented with a series of different tones. Most of the tones are the same pitch, but occasionally a tone is presented at a higher pitch (the oddball or target tone). Participants have to press a button as quickly as possible when they hear that higher-pitched tone, but not press it for any other tone. Sometimes we also play a few funny, random tones, just to mix things up a bit. The latter tones examine the brain's response to novelty.

The Oddball Task has been around for over fifty years. It turns out that the target tones elicit a very large and beautiful electrical brain response. The most prominent component of the target ERP is the P3 (the third positive peak following the onset of the target tone). The P3 is a very complex waveform; for years I attended an annual meeting where the only topic was what scientists thought this component meant. Since its discovery in 1965[3] there have been thousands of scientific papers published about the P3 component of the ERP.

Why is this interesting? The P3 component has a lot of clinical utility for researchers. Most forms of mental illness are associated with abnormalities in the P3. In schizophrenia, the P3 is reduced in amplitude and delayed in latency, effects that are most prominent in sites over the left temporal lobe of the brain. In serious depression, the P3 is reduced in amplitude over the frontal lobes of the brain. So variations in the amplitude (height), latency (duration), and topography (shape) of the P3 can help to identify brain processes related to mental illnesses.

After all these scientific meetings and publications, do we know what the P3 means? The interpretation I favour is that the P3 is a probe for how the brain is functioning; when the P3 is distorted, it tells us something is wrong in the brain.

The brain tends to respond automatically, even reflexively, when processing those oddball or salient target stimuli. There is an automatic orienting process that occurs. It's like the brain says, 'Ahh, that's an important stimulus. Let's make sure we process it.' Let me give you an of what I mean here.

One summer I was hiking with my fellow whale research colleague Michael on the Pacific Crest Trail in California. We were hiking just south of Yosemite National Park. We had been out for over a week and we were trying to finish our hike by exiting at the Kings Canyon National Park, some ten miles away. My two black female German shepherds, Andi (short for Andes mountains) and Alaya (short for Himalaya mountains), were along for the trek. Andi and I were in front of our group, Alaya behind me, with Michael pulling up the rear. As we were hiking along, Andi and I heard a funny noise coming from the trail ahead of us. It sounded like a bird fluttering under a rock. Andi's ears perked up, and I recall thinking to myself at the time that we had both just had a P3 novelty response.

Andi proceeded to go over and investigate the rock from where the sound was emanating. When she was about a foot away, she froze and then leaped straight back up in the air. Attached to her backpack was a baby rattlesnake. I quickly flicked the snake off Andi's backpack with my trekking pole, and we watched it slither away down the hill. We just looked at each other. We all had big P3s as a result of that baby snake, but everyone was okay. It turned out that the baby snake's tail was not fully formed, and it did not make the distinctive rattle with which we were all familiar.

About a quarter mile later we heard the same sound again. However, this time our brain had prepared us for the target sound. All of us, canines and humans, immediately jumped back out of the way. P3 again.

We can translate this experience into the laboratory by simply requiring participants to press buttons for auditory stimuli we designate as important or salient. A robust P3 can be elicited by the salient stimuli.

Why does the brain respond like this? Well, it's adaptive to respond quickly to important stimuli in our environment. If our brain did not permit us to learn quickly, Andi and I might have been killed by that baby rattlesnake, and our bodies would be decomposing on the side of a mountain trail.

Scientists believe that the brain has a reflex that engages anytime a stimulus is potentially important. By engaging this brain reflex, we are preparing our minds to process and adapt to important stimuli. This is the essence of the P3.

My laboratory has shown that the Oddball Task elicits activity from more than thirty-five regions of the brain. The synchronous activity of these thirty-five or so regions results in the waveform we record on the scalp as the P3. Any abnormalities in the thirty-five regions of the brain responsible for the P3 will result in alterations in the amplitude, latency, or topography of the component. When we find that a mental health disorder is associated with some alteration in the P3, we have to figure out what brain regions might be causing the abnormality. In this way the P3 can serve as a probe for what's going wrong in these thirty-five or so regions of the brain.[4]

I had Shock Richie complete the Oddball Task, and I took the data home for processing. As I analysed the brain wave data that night, I noticed something very odd about Richie's P3. Not only was it reduced in amplitude relative to other (nonpsychopathic) inmates over the front part of the brain, but it went hugely negative right after the P3. I stared at that waveform for a long time that evening.

I collected Oddball Task data from forty more psychopaths in addition to Shock Richie. I also collected brain wave data from forty inmates who scored low on the Psychopathy Checklist. The latter subjects constitute the nonpsychopathic control group (see Figure 2).

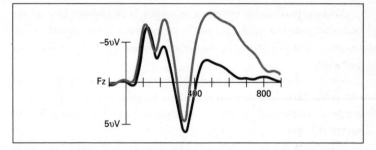

Figure 2. Event-related brain response (ERP) from a frontal brain site for forty psychopaths (grey line) compared to forty nonpsychopaths (black line) for the auditory oddball stimuli. Note the prominent difference between the psychopaths and nonpsychopaths starting at about 400 milliseconds and extending out to 800 milliseconds. This is the abnormal brain wave response of psychopaths. Units on the x-axis are in microvolts with negative amplitude plotted up; units on the y-axis are in milliseconds following the onset of the oddball stimuli. Data from Kiehl, K.A., et al. (2006). Brain potentials implicate temporal lobe abnormalities in criminal psychopaths. *Journal of Abnormal Psychology*, 115, 443–453.

I printed out the eighty-one inmates' brain responses to the Oddball Task. I assigned a random number to each case and removed any details about whether the plot was a psychopath or not. I asked a research assistant to sort the brain wave plots based on the presence or absence of the weird P3. The assistant correctly sorted forty out of forty-one psychopaths. None of the nonpsychopaths were put in the wrong group. In other words, the weird P3 was literally diagnostic of psychopathy. Ninety-seven percent of the psychopaths showed this weird brain wave response, and none of the nonpsychopaths showed any evidence of it. It was a fascinating result.

Over the course of the next year, my lab collected hundreds of additional control subjects who were not in prison, studying their oddball responses to see if we could find any indications of the weird P3 in anyone other than a psychopath.

One morning an excited research tech came into the lab to show me a plot of a brain wave response he had recently collected. I assumed it was from a new psychopath since this tech had just been trained to collect ERPs at the prison. I reviewed the plot and concluded that it matched the template we had created for the psychopathic brain. I congratulated him on his first psychopathic brain wave.

All at once the blood drained from the tech's face.

'What is the matter?' I asked.

'Y – you don't understand,' he stammered. 'That plot isn't from an inmate. It's from my roommate.'

My first thought was my assistant better get a new roommate as soon as possible. My second thought was that we couldn't disclose anything about this plot to his roommate; it would be a serious breach of ethics.

My assistant was shaking. He started babbling about the behaviour of his roommate he had met on the classified-advertising website Craigslist. The guy seemed nice enough, was very talkative, and, although he smoked a lot of marijuana, there didn't seem to be anything wrong with him on the surface.

In the end, my assistant decided that it was probably a good idea for him to move out. That's the best way to avoid being a victim of a psychopath – trust your gut.

A couple weeks later my assistant told me he had some interesting news. First, he had moved back in with his parents and felt a lot better. But he said that before he had moved out, his roommate had confided in him that his father had been convicted of murder.

Interesting. Would our quest to figure out the psychopath's weird P3 response lead us to conclude there is a genetic component to the abnormalities in the brain?

chapter 5

The Psychopath Magnetized

ғact: psychopaths are six times more
likely than other criminals to commit
new crimes following release from
prison.

Determined to figure out the psychopaths' weird P3
brain riddle, I studied thousands of scientific articles published on
the brain wave known as the P3. In parallel, I presented my unusual P3 findings on psychopaths at scientific meetings. I designed
a poster describing the results and put the brain waves up in big coloured graph plots so everyone could see. But nobody had a clue what
the weird P3 meant.

Next, I printed out a smaller version of my poster and went to
conferences to ask other scientists who studied ERPs to take a look at
my plots. Again, nothing.

I started taking my data to other scientists' presentations, asking
them to look at my plots. I made friends with PhD students from
all the laboratories that did ERP studies. None of them had ever
seen anything like my weird P3 results. In fact, most of the graduate
students initially thought my results were artefacts. But when I told
them about the consistency of the results, they would often ask to see
the plots again and would just stare at them for a while. Then they
would look up and apologize; nobody recognized the weird brain
wave.

After about a year of taking my plots around with me and finding no answers, I was getting frustrated. And I was a little scared

that I might not get my PhD if I did not have a solution to this riddle.

Then I received an invitation to give a lecture at a conference in Budapest, Hungary. Maybe somebody in Europe could help solve the riddle of the psychopaths' odd brain waves. I accepted the invitation and flew over to give my talk on the psychopaths' weird P3 responses.

The audience found my results intriguing, but nobody had an answer about what they might mean. In attendance at that conference was Dr Robert Knight, a neurologist whom I knew from my UC Davis days. Bob had published studies showing that the P3 is abnormal in patients with brain damage to the frontal lobes. But the abnormality wasn't the same as what we were seeing in psychopaths. Nevertheless, I asked Bob about my psychopathic plots.

We went to an outdoor bar that overlooked the city and ordered a couple beers. Bob suggested I apply for a faculty position at the University of California, Berkeley, where he had just relocated from UC Davis. I was very interested in the faculty position, but I told him that I wasn't likely to receive my PhD anytime soon if I didn't figure out a good interpretation of the strange P3 in psychopaths.

'Let me see that plot again,' he said.

He stared at it for a while. I had been trained to use the same software Bob used, so he was very familiar with the custom plots and the layout of my findings. After a while, he looked up at me and said, 'I think I have seen this before. But I don't remember where.' He dropped the plot back on the table. 'I'll send you a list of my papers. You can look through them to see if this plot shows up in any of them.'

Bob sent me the list of more than two hundred papers he had published.

I went to the library and spent a fortune to photocopy all his papers. I scoured every paper. Nothing. I double-checked and, again, found nothing. I became an expert in Bob Knight's career.

I e-mailed Bob and told him that I did not find anything similar to my plots, but thanked him for sending me the list of all his publications. He wrote back that I should check the book chapters he had written. Sometimes his group published brain wave results in them.

Scientists are usually concerned only with peer-reviewed publications, where two or more scientists critique manuscripts anonymously and provide feedback to help the journal editor decide to accept or reject the scientist's manuscript. Most good journals publish only 10 to 20 percent of the articles they receive, so it's quite difficult to get a manuscript published by peer review. Book chapters, on the other hand, are often invited by the person organizing the book, and they are typically not peer reviewed. In academia, book chapters aren't viewed as highly as peer-reviewed articles.

Bob had authored over fifty book chapters. Back to the library I went, to spend another fortune on photocopies.

The list of book chapters was organized alphabetically by the surname of the lead author. I placed the photocopies of the chapters in a stack on my kitchen table in the same order. As I ate dinner, I flipped through the chapters one by one, looking at the brain wave plots. As I worked my way through the pile of book chapters, I slowly drained a bottle of wine.

I got to the last book chapter, written by Yamaguchi and Knight, published in 1993.[1] It was a study of the P3 in patients who had brain damage to the lateral and medial temporal lobe. I turned the page and when I looked down on the plot before me, I gasped (see Figure 3).

There it was. Patients with brain damage to the lateral and medial temporal lobe have the same weird P3 response as psychopaths. I could not believe it. I memorized the chapter.

From time to time I see Bob at conferences. At a recent meeting he was giving a lecture when someone in the audience asked him about the results from one of his older papers. He looked over at me and said to the audience: 'I don't remember, but Kent Kiehl does.'

The audience laughed, and I took the microphone Bob handed me and told them of the results from the 1993 paper by Yamaguchi and Knight.

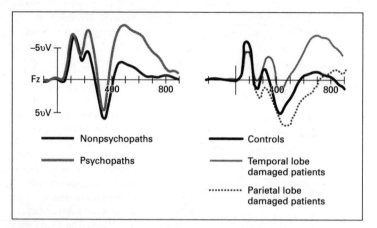

Figure 3. Brain wave plots for psychopaths from Kiehl et al. (2005; left plot) and Yamaguchi and Knight (1993; right plot) for auditory oddball stimuli. Both plots have been adapted to the same time scale and amplitude. Negative amplitude is plotted up. Note the similarities between the psychopaths' brain waves and those of patients with temporal lobe damage. Both groups show enlarged negative response at 200 milliseconds, reduced positive response at 350 milliseconds, and an enlarged negative response at 400–800 milliseconds compared to control subjects.

A Window onto the Psychopathic Mind

Armed with my answer to the riddle of the psychopaths' weird P3 brain waves, I was feeling confident about the next set of studies I wanted to do with them to further elucidate what might be going wrong inside their brains.

Most of the studies I designed sought to examine whether the medial and lateral aspects of the temporal lobe were abnormal in psychopaths. These regions include the amygdala, the hippocampus, and the temporal pole (the full name is anterior superior temporal gyrus; see Figure 4).

The amygdala is an almond-shaped region deep in the brain that is responsible for amplifying salient information. The amplification of salient information interrupts ongoing thinking and makes us pay attention. For example, you are walking along a crowded street

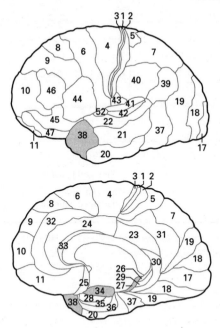

Figure 4. Renderings of the lateral (top) and medial (bottom) of the human brain. The medial view is as if you sliced the brain down the middle and pulled the two halves apart to look inside. The numbers represent areas defined by a labelling system developed by anatomist Korbinian Brodmann in 1909. Scientists use the Brodmann numbering system to help them compare results across studies and across laboratories. The results from the weird P3 brain wave in psychopaths implicated the amygdala (Brodmann area 34), the hippocampus (Brodmann area 27), and the temporal pole (Brodmann area 38).

and you hear a loud bang and quickly turn to identify the cause and location of the noise. This noise has just been amplified by the amygdala, causing a startle reflex, alerting you that it might be critical to pay attention to what's going to happen next. The startle response requires the amygdala.

The amygdala also helps us learn what stimuli are important to amplify and raises this information into awareness; for example, learning that a hot stove is not something to be touched or that an electrical socket is not something to lick with your tongue. The amygdala assists learning these basic fear and emotion contingencies.

The hippocampus is the seat of memory in the brain. It's responsible for the consolidation and storage of memories. It's one area of the brain that continues to grow throughout life, becoming thicker as one ages.[2] The hippocampus is particularly good at storing emotional memories.

The temporal pole is a bit of an outlier here. The amygdala and hippocampus are classic members of the limbic system, which is generally considered to be responsible for the control of affective and emotional processes in the brain. But the temporal pole was not part of the limbic system first described by neuroanatomists Paul Broca (1824–1880) and James Papez (1883–1958). The temporal pole is an area of what is called the *heteromodal association cortex*. This means it is a region where lots of sensory information arrives and is integrated. So auditory information and visual information converge at the temporal pole and are merged for subsequent higher processing, similar to how filmmakers integrate sound and video to make movies.

Studies have shown that brain damage to the right temporal pole can lead to impairment in prosody of speech.[3] *Prosody* is the affective intonation in speech. Individuals who suffer damage to the temporal pole can't describe, or have problems indicating, what type of emotion is being conveyed by affective speech. Similarly, damage to the temporal pole may cause impairments in understanding abstract representations of speech, like metaphors.[4]

My literature review on the functions of the amygdala, hippocampus, and temporal pole led to a number of new lines of research. Some of these new studies continued to use brain wave recordings. But what was really exciting were the studies we had been planning that used a new technique called *functional magnetic resonance imaging*, or fMRI.

MRIs and the Psychopathic Brain

MRIs, or magnetic resonance imaging devices, use a combination of strong magnetic fields and radio waves to create amazing images of human anatomy. MRIs have been around since the mid-1980s and today are present in many hospitals around the world. MRIs do not use radiation or x-rays and are thus considered noninvasive and very safe for research purposes.

In addition to creating beautiful pictures of brain anatomy, the latest advances in MRI permit scientists to study the brain in action.

The most common technique in the neuroscientist's arsenal in this regard uses a souped-up MRI system to quantify changes in levels of oxygen-labelled blood in the brain. Just like muscles, neurons in the brain need oxygen when they are working. The lungs function to attach oxygen to the haemoglobin molecule in blood, and the artery system transports oxygen-loaded haemoglobin to the brain (and to muscles). Blood with a lot of oxygen in it has a different MRI signal than blood with less oxygen in it. Blood with a lot of oxygen is a bright red colour (arterial blood), and blood with less oxygen is blue (venous blood). The MRI scanner can be tuned to record snapshot images that map the precise location in the brain where oxygen is being delivered and consumed (i.e., changing from red to blue in colour). Over the course of a few minutes, scientists can determine which brain regions are consuming oxygen while participants are doing specific tasks in the MRI scanner. This is known as *blood oxygen level dependent (BOLD) imaging*. We typically refer to this technique as functional MRI, or fMRI.

Functional MRI was discovered in 1992[5] while I was still at UC Davis. I was working for Professors Mike Gazzaniga and Ron Mangun, who both quickly started using fMRI in their laboratories. The amazing research environment at UC Davis allowed me to immerse myself in the technical aspects of MRI from the very beginning of the field.

When I left for graduate school in the summer of 1994, I had been working with fMRI data for over two years. As I settled into my new home in Vancouver, I started searching for the best MRI system in the city where we might be able to do fMRI studies of psychopaths. The BOLD fMRI technique places a lot of demand on the MRI system since it pushes the hardware to its max when collecting functional brain imaging data. In a normal MRI session, about ten or twenty images will be collected. But in fMRI, ten or twenty *thousand* images will be collected. In addition to the special hardware on the MRI, the room needs to be equipped with high-resolution projection systems, custom video screens, fibre-optic response devices, and special communication cables that all have to be MRI compatible. 'MRI compatible', I often muse, is a euphemism for money. For example, a regular joystick for a video game might cost $20, but a fibre-optic MRI-compatible joystick typically costs $2,000. Thus, it takes a lot of customization and money to make fMRI possible on plain-vanilla MRI systems.

My search revealed that the University of British Columbia Hospital had just purchased a brand-new General Electric (GE) 1.5T MRI system. I called around until I found out who was in charge of the new system. Dr Bruce Forster, a radiologist, was the man I needed to meet.

I started my reconnaissance of the University Hospital by circumnavigating the grounds searching for the MRI bay. I found that the MRI bay had been built right into the side of the hospital and had its own brick service road. I assumed the brick road had been built to facilitate delivery of the MRI system. Clinical MRI scanners can weigh up to sixty thousand pounds and require special installations.

I entered the University Hospital through the basement cafeteria and looked for signs to the MRI area. The signs led down a long corridor to the main entrance of the MRI suite. I approached the receptionist, introduced myself, and asked if I could speak with Dr Forster. The receptionist noted that Dr Forster's office was upstairs, but he was just preparing to leave for the day so I had better hurry if I wanted to catch him.

I sprinted down the corridor and up the stairs. Two floors up, I opened the door and briskly walked down past door after door of radiology offices. It was about 6 p.m. and one of the offices at the end of the corridor had its door open and its lights still on.

As I approached the door, a man exited the office, and I walked right into him. I managed to mutter a quick apology. Recognizing I was a stranger, he asked if he could help me find my way. I had completely ignored the sign on the door at the end of the corridor that said AUTHORIZED PERSONNEL ONLY.

'Yes. I am looking for Dr Forster.'

'That's me. What can I help you with?' he answered.

'I'm a graduate student here at UBC studying criminal psychopaths. I want to know if I can bring them out of the maximum-security prison and scan them on your new MRI,' I said flatly.

He coughed, took a step back, and reevaluated the situation.

Dr Forster was dressed perfectly; even his socks matched his suit. He had a flawless goatee, and his hair was brushed back along his scalp. He looked like a contemporary Sigmund Freud.

When he finally spoke, in a very deep yet softly commanding voice, he said, 'Let's sit down for a minute so you can tell me more about what you are thinking.'

He pointed to the chairs in his immaculately clean office. I walked over and sat down. He positioned himself behind the desk, closest to the door.

'So tell me again what you are interested in doing,' he asked.

I explained my research and that of my supervisor, Dr Hare. I told him about my training in fMRI with Drs Gazzaniga and Mangun at UC Davis. Another graduate student in Dr Hare's lab and I were planning to ask the Canadian Department of Corrections whether they would transport inmates from the prison to the hospital so we could scan their brains. I wanted to know if his MRI scanner was capable of doing fMRI.

He leaned back in his chair and told me that never in his life had he expected someone would ask him to use his MRI scanner to scan criminal psychopaths.

'Tell me more,' he asked.

After a thirty-minute conversation, Dr Forster took me back downstairs to the MRI suite.

We headed into the MRI bay and he introduced me to the lead physicist, Dr Alex Mackay. Dr Mackay headed up the research group studying white matter lesions in the brain. The MRI was very busy with clinical patients during the day, but Mackay had a deal to do research every Wednesday night from 6 p.m. to midnight or later.

I gave Dr Mackay a quick rundown of my background and fMRI training.

'Oh,' he said. 'We built this MRI with the best hardware just in case someone wanted to do fMRI.'

Dr Forster chimed in that he would love to be the first group in Canada to use fMRI for presurgical mapping.

Presurgical mapping refers to a procedure that brain surgeons complete prior to removing a tumour and adjacent brain tissue. Brain surgeons want to remove any bad tissue without removing any tissue that is critical or eloquent. *Eloquent cortex* refers to areas of the brain that control things such as your ability to speak or move your tongue. The old way to do presurgical mapping is to remove the skull and electrically stimulate the brain in order to figure out what bits of the brain do what. Surgeons painstakingly map out with electrical recording devices, not that different from EEG, which brain regions are eloquent before they go in and remove tumours and adjacent tissue that might also be affected.

But fMRI can do this presurgical mapping without using a scalpel. Using an MRI, scientists can map out the regions involved with language and motor functions. The brain surgeon is then given a map of the patient's eloquent cortex without any need to remove the skull and resort to electrical stimulation.

The University of British Columbia is also the home of Dr Juhn Wada, who pioneered a technique to put one hemisphere of the brain or the other to sleep with anaesthesia. In this way surgeons could work out what side of the brain contained language. In most people, language is in the left hemisphere. But in a few people – about 10 percent of the population – language is handled in the right hemisphere. It's safer to remove brain tissue from the nonlanguage hemisphere when resecting tumours. The Wada technique helps to localize which hemisphere has language in it. But the Wada technique is invasive, and sometimes people die from the procedure before they get into brain surgery to remove tumours. Replacing the Wada technique with noninvasive fMRI would be very valuable.

So I volunteered to do the presurgical mapping for Dr Forster. Dr Mackay also generously allowed me to volunteer to work with his group on MRI studies of white matter lesions. We all agreed that we would work together to find way to get prisoners scanned on the brand-new University Hospital MRI.

I was surprised that Dr Forster didn't run out of his office when I cavalierly told him the plan to transport maximum-security inmates eighty miles from a prison to the university and then remove their handcuffs and put them in the MRI. I may have forgotten to tell Dr Forster we have to take their handcuffs off when we scan them because no metal is allowed in the MRI room. I figured we could cross that bridge when we got there. I'd put enough new things on his plate.

Over the next several years, I spent nearly every Wednesday night from 6 p.m. to midnight working for Dr Mackay. I was often the volunteer, a guinea pig, who got in the MRI scanner as a test case as Alex and his team of PhD students developed new MRI pulse sequences. My brain was scanned dozens of times in those years. I would come to find the MRI bore a comfortable place to sleep. Indeed, it is hard for me to not fall asleep as soon as I get into an MRI at this point.

We managed to do a couple of presurgical mapping patients for Dr Forster. He was thrilled and presented the data at a number of professional meetings, discussing the future possibilities of what this new technology might bring to bear on such cases.

One of the cases was an eighteen-year-old girl who had a small malformation in the arteries and veins in her brain. About a week before I met with her, she had started experiencing symptoms that included tingling in her face, hands, and tongue. Then one day she woke up and could not speak, although she remained conscious and alert. She had walked to her parents' bedroom and tried to communicate to them what she was experiencing but she passed out. Her frantic parents rushed her to the UBC emergency room, where Dr Forster found the abnormality. He suggested she undergo the presurgical mapping fMRI.

I designed tasks to map out the hand, face, and tongue areas of her brain. Then I designed a task to reveal the speech producing regions of the brain. On the day of her scan, I met with her and helped her practise all the tasks. She was very curious about the procedure, but her eyes relayed her emotion and concern. She was worried about what we might find.

We learned where the arteries and veins had become twisted and which parts of the brain were linked to the area that the surgeon was going to glue shut. The procedure literally uses glue to close the faulty artery. But the brain tissue supplied by that artery could die as well, removing her ability to move her face, hands, or tongue. The procedure could take away her ability to speak.

I agonized over the analyses of her brain imaging data. I had to make sure there were no mistakes. I did not want a surgeon removing an important piece of her brain because I made a mistake. I reviewed the results with the neurosurgeon and told him about all the possible caveats and limitations we know about the fMRI technique.

The surgeon used our results to help guide where he would glue shut the faulty artery. To everyone's delight, she made a full recovery.

fMRI Task Development

I collected brain imaging data on weekends when I was not in prison. Most graduate students were happy to volunteer for MRI studies in return for getting pictures of their brains. Before the end of our first year of scanning, nearly all the PhD students at UBC had received MRI pictures of their brains. Our team was developing all the tasks and procedures we were going to perform on the inmates once we arranged to transport them for scanning. I spent long days at the MRI suite collecting and analysing data, and even longer days at the prison interviewing inmates and collecting more brain wave studies.

I could not have coordinated the logistics of the first fMRI study of criminal psychopaths without a lot of help. Andra Smith, a PhD student at UBC in neuroscience, had started graduate school studying aggression in rats but then developed a severe allergy to them. So she switched to work with prisoners and study psychopathy. Andra and I spent a lot of days working together at the prison; she was the one who wrote most of the grants that we got to support our project. None of the original brain imaging studies of psychopaths would have been completed without her help and dedication.

Dr Peter Liddle, a physicist and psychiatrist, was instrumental in training us in the intricacies of brain imaging analyses. Before moving to UBC, Dr Liddle had been a founding member of a group at Hammersmith Hospital in London, where the software package called *Statistical Parametric Mapping*, or SPM, was born. Peter had hired a young psychiatrist caled Karl Friston to help do clinical assessments of patients with schizophrenia. Karl also had an interest in maths, and he went on to develop the most popular software in the world for the analysis of brain imaging data. Over 80 percent of the peer-reviewed publications in the world use SPM to analyse their imaging data. I was fortunate to be able to work with Peter, Karl, and their respective laboratories and learn the best ways to analyse brain imaging data from the very beginning of the field.

UBC's new GE MRI machine was not without its quirks. We were pressing the machine to the limits, and GE engineers had not anticipated some of the problems that would arise. One challenge was that the system was limited to being able to collect only 512 images consecutively. But in fMRI we collect thousands of images at a time. In my quest to solve this problem, I met a couple of fellow scientists who were trying to overcome this and other issues with the GE scanner.

At an MRI meeting in Vancouver that year, I met a physics graduate student from Wisconsin called Bryan Mock. Over a burger and a couple of beers, he transformed how we collected data on our GE scanner. Bryan provided us with code to pull the MRI data off the scanner and reconstruct the raw data on another computer. It was Bryan's 'offline reconstruction' that allowed us to circumvent GE's 512-image limitation. Bryan then told me if we wanted to make the scanner go faster, we could turn off a couple of the heating calculations. It appears that GE engineers never anticipated scientists pushing the MRI system as hard as we do when collecting fMRI data. GE had installed a little governor, like a system that limits how fast a car can go, on the MRI. Bryan taught me how to turn off the governor.

After my lunch with Bryan, I called the techs in the MRI suite and had them turn off the governor on the GE MRI. Then I asked them to run our fMRI sequence on a phantom (the water bottle dummy we use in MRI to test sequences). The techs dropped the phantom in and ran the new sequence. Our MRI was 40 percent faster! Amazing.

Bryan was quickly hired by GE right out of graduate school. He now runs the MRI product development team, and I continue to buy him burgers and beers anytime we get together.

I also met by e-mail a graduate student at the University of Maryland, Vince Calhoun, who was trying to solve the same problem of lack of speed on the GE scanners he was working on. Vince was an engineer working on novel ways to analyse brain networks. So I shared with Vince all the tricks I learned from Bryan Mock. Years later Vince Calhoun and I became junior faculty together. And Vince continues to buy me burgers and beers.

My fellow graduate student Andra Smith took to dealing coffee from my Starbucks source to the Regional Escort Team of the Canadian Department of Corrections. The Escort team is responsible for transporting inmates from prison to prison or from prison to special medical appointments. After about six months of supplying the team with free coffee, Andra approached them to see if they would transport the inmates from our prison out to the University Hospital for MRI scans. They said they would be happy to help.

The prison security staff was a little wary about letting the inmates know what day they were going to be transported to the hospital for MRI research. The Regional Escort Team wanted to make sure that the inmates did not try to plan an escape. The team especially wanted to make sure the inmates did not recruit confederates on the outside to help break them out when they were being transported. Our project had to be top secret.

Escape was a pretty serious concern. On June 18, 1990, confederates had hijacked a helicopter and landed it inside the fences of a maximum-security Canadian prison (coincidentally named Kent Institution). An inmate serving a life sentence had organized the escape, and another inmate joined him when the helicopter landed inside the prison walls. The confederates shot and seriously wounded a guard. The escape plan was right out of a Hollywood film. The inmates and the confederates made it out of the prison but were caught a few weeks later hiding out in the woods.

One day I was talking with the head of the Escort Team about security measures, and I asked him whether they prevented inmates from making phone calls on days we transported inmates for research. He told me that on transport days, the prison planned to cancel all phone privileges to make sure that an inmate did not call someone outside and tell him or her the inmates were being transported.

That should help, I thought. But then I asked the team leader what if the absence of a call was the signal? In other words, if an inmate called the same person every day and then one day failed to call, that might be the clue it was the day the inmates were being transported.

The team leader hadn't anticipated that. So for the weeks leading up to our first scan day, the prison randomly shut down phone privileges for the entire prison. The Canadian Department of Corrections was fabulous; they were doing all this for research.

One morning in the middle of the summer we began our brain scan project. Five psychopaths were transported under armed guard eighty miles from Abbottsford to Vancouver. There were four vehicles in total. The first vehicle was a scout, and it travelled ten minutes ahead of the caravan to make sure the coast was clear. Then came the security car, followed immediately behind by the transport van. The van looked like an armoured car, similar to those used to haul money. Five minutes behind the van was the trail car with two heavily armed guards.

The University Hospital set up a special traffic route and posted guards around the hospital. The university even built a wooden shield that closed and locked the van full of inmates down in the MRI delivery area. The van was able to drive down the brick road and park within ten feet of the MRI exterior entrance.

I'll never forget our first delivery.

I'd awoken before dawn and got busy prepping for the day. We had the MRI suite cleared of anything that the inmates might use for a weapon. We had hospital security guards checking the perimeter to make sure there were no confederates lurking nearby who might try to help spring an inmate.

I had picked up a couple dozen doughnuts and lots of coffee for the guards and inmates on my way into the hospital. Dr Liddle and Dr Forster were there, as well as my fellow PhD student Andra Smith and our trusted MRI tech Trudy Shaw.

The scout car arrived at 7 a.m., well before most of the hospital staff had reported for work. The guards came in and looked around, double-checking our local security measures. They called on their radio for the van to descend into our little fabricated car park. After the van came to a stop and the guards secured the wooden doors, they unbolted the huge metal arm that locked down the rear door of the armoured vehicle. The inmates were seated on steel benches shackled to the floor. Each inmate was unshackled from the floor and escorted into the hospital. The inmates squinted as they adjusted their eyes from the dark confines of the van to the sparkling MRI suite. Shackles around their hands and feet were both attached to a chain around their waist. The inmates shuffle-walked into the waiting room and sat down in the plush seats. The guards removed the belt chain so the inmates could raise their hands and eat their doughnuts and drink some coffee. The shackles were left on their hands and feet.

The first inmate to be scanned was my pal Shock Richie. He was escorted into the MRI room for prep. I explained the procedures to Richie and gave him examples of the tasks he was going to perform while in the MRI scanner.

Shock Richie said, 'You'll let me know if there is anything wrong with my brain?'

'Don't be shocked if there isn't anything in there,' one of the guards quipped.

Apparently, Richie had already given the guards a bit of trouble when he showed up naked at his cell door when they called him out at 5 a.m. for the transport. The guards had waited, unimpressed, while Richie got dressed.

Shock Richie smiled at me as the guard knelt down to remove the shackles around his feet. As the final set of cuffs came off, Richie rubbed his wrists for a minute as we walked into the MRI room. I noticed Dr Forster watching intently from the window in the control room. He had come to grips with the fact that the inmates were going to have all shackles removed before they went into the MRI room. But I never should have told him to watch the movie *Con Air* with Nicolas Cage and John Malkovich, in which a group of convicts escape their shackles and take over the transport plane; he had nightmares for weeks.

Shock Richie jumped up on the MRI table. We positioned him on the table and made sure that he was comfortable. Trudy handed Richie the pneumatic squeeze ball, explaining to Richie that the little ball sets off an alarm in the control room and is to be used only in case of emergencies. She cautioned that he was to squeeze the ball only if he needed to get out of the MRI. I knew what was going to happen next even before it occurred.

Shock Richie squeezed the emergency ball. A piercing alarm went off in the control room, startling Dr Forster, who spilled his coffee on his suit. There was a loud commotion and noise coming from the control room as the group struggled to disengage the alarm.

Richie, able to see the commotion he had caused, smiled; he told me to make sure that everyone knew his name was 'Shock Richie'.

Trudy spent a few more minutes getting Richie settled; I showed him the video screen he would be watching while doing our tasks. Trudy and I then exited the MRI room, and I pulled closed the heavy, magnetically shielded door behind us. The click of the lock elicited a feeling similar to that of the lock shot firing at the prison doors, but this time I was locking my first psychopath into the MRI.

Trudy was laughing a bit at Shock Richie's antics. Everyone was amped up that morning, and the little joke Richie had played eased the tension in the air.

We sat down at the MRI console and started our brain imaging protocol. We talked to Richie using the intercom system; he was ready to go.

The MRI started its familiar beeping as the magnetic fields collided and the radio waves flipped protons about. After a few minutes, the first images of a criminal psychopath's brain materialized on the computer screen.

I stared at that image, anticipating there was going to be some huge hole in Shock Richie's brain. Dr Forster appeared over my shoulder and studied the image with me. He reached past me and flicked the dials, and the image spun through a whole series of slices of the brain from top to bottom, then left to right. I directed all my attention to making sure Shock Richie wasn't climbing out of the MRI and trying to escape. Fortunately, we could see Shock Richie the entire time from our vantage point in the MRI control room, and there was only one way in and out of the room.

Dr Forster said Shock Richie's brain looked pretty normal. The guard sighed in disbelief.

I knew Dr Forster's assessment meant that Shock Richie's psychopathic behaviour wasn't going to be explained by some tumour or other gross brain abnormality. The computer algorithms were going to have to analyse his brain very carefully.

To his credit, Shock Richie did everything that he was asked to do. He performed all the tasks well, didn't move his head much, and the images looked great. As he was removed from the MRI, he sauntered over and had a seat in the chair of the MRI control room. The guard applied his leg and wrist shackles.

I let Shock Richie have a look at his brain on the computer screen.

'Surprised it's in there?' I asked.

'Yes, but not shocked,' replied a smiling Richie. 'Happy to see something in there. You'll let me know if I have the best brain of the bunch, won't you?'

He smirked and looked at me with those empty eyes.

I had to figure out what was different about Shock Richie's brain, what was behind that cold, flat, unemotional look in his eyes.

The rest of the day went smoothly. Each of the remaining inmates got scanned, and none squeezed the emergency ball. I gave each inmate a printed picture of his brain. The inmates compared them to one another like little kids. Shock Richie told the rest of the inmates that his brain was the best. He pointed to his thick corpus callosum and highlighted it for the rest of the inmates. I had given the inmates a quick anatomy lesson as I handed out their pictures. Richie had discovered that the fibre bridge that connects the two hemispheres of the brain together, known as the *corpus callosum*, was unusually thick in his picture.

The Regional Escort Team and inmates alike made jokes about Richie's dense brain structure for the rest of the day. We fed everyone pizza for lunch, and the caravan departed just before 3 p.m. after a highly successful scan session.

As I watched the armoured van creep up the little brick road on its way back to prison, I felt the adrenaline rush that had lasted the entire day start to subside. My body was telling me I didn't have much left in the tank, but I wanted to get the computers cranking on processing and backing up all the brain data.

Retreating to the laboratory, I initiated a script I had coded to analyse the data overnight, and then I walked to the car park to head home.

Over the course of the next couple of years, we arranged ten more MRI sessions, scanning over fifty maximum-security inmates. Each of the visits went well, with only minor complications. I gratefully acknowledge the efforts of the Regional Escort Team and the rest of the Canadian Department of Corrections, the University Hospital staff, and especially the MRI technologists and my fellow lab members.

The First Results

In the two years prior to scanning Shock Richie, we developed the tasks that we would use to examine brain dysfunction in psychopaths. In particular, we wanted to develop tasks that engaged the limbic system in the brain, the system mostly associated with emotional processing. Given the profound deficits in emotional behaviour that typify psychopaths, we were primarily interested in probing that system with our first brain imaging studies.

The parts of the limbic system we were most interested in included the little almond-shaped region known as the *amygdala*. I call it the brain's little amplifier. The amygdala was hypothesized to be a critical node of impairment in psychopaths. Other critical areas we scanned included the anterior and posterior parts of the cingulate cortex, regions believed to be related to attending to emotional parts of language and other stimuli.

A favourite task to emerge from our pilot studies was an emotional memory paradigm. In this task, participants are asked to memorize a list of twelve words, presented one at a time on the screen. The *encoding phase* was followed by a twenty-second *rehearse phase* where participants ruminate on the words just presented. Then a *test phase* is given where twelve words are presented, again one at a time. Half the words in the *test phase* are from the prior *encoding phase*, and the other half are new words that participants have not seen before. Using a button box, participants have to indicate whether the word is from the previous list or not. About twelve different lists are presented over the course of about fifteen minutes.

What we did not tell the subjects was that the lists were composed of either emotional or neutral words. Examples of emotional words were things like 'hate', 'kill', or 'death'. Neutral words included words like 'table', 'chair', or 'leg'. We found that normal people recall emotional words better than neutral words, and that the amygdala and anterior and posterior cingulate are more engaged when processing emotional words than when processing neutral words. It's amazing what fMRI can show you about how the brain works.

I stared at the huge computer monitor as the little red bar slowly climbed to the top of the graph, indicating that the data analysis was nearing 100 percent completion. I had just about finished processing the first study comparing psychopaths and nonpsychopaths on the emotional memory task. It had taken four prison transports to collect enough inmates for the study. I had spent hours making sure that the data were perfect, performing all the postprocessing steps for optimized analyses. I'd had to exclude a few inmates because they moved their heads too much, which made the images too blurry to analyse. I'd double- and triple-checked all the output to make sure the results about to pop on the screen were valid.

I sat there sipping my coffee and waiting. It seemed to take fo ever for the bar to climb to the top and hit that last mark.

Finally, *Finished* popped up on the top of the screen. I jumped on the keyboard and typed in the comparison I wanted to view.

Where do psychopaths' brains show deficits in emotional processing?

The computer started to hum as the rendering of brain results strained the processor and memory capacities of the Sun workstation. As the image of the first ever results of a brain imaging study of criminal psychopaths started to appear, I thought about the road it took to get here: all the days and weekends spent in prison, all the commutes back and forth, all the interviews with inmates, all the meetings with administrators from the University Hospital and the Canadian Department of Corrections, all the planning, all the years.

The screen filled with a large image depicting four views of the brain, with blue colours indicating where psychopaths showed deficits in emotional processing. It took my breath away.

The amygdala and the anterior and posterior cingulate were lit up in bright blue – meaning that they were less active in psychopaths than in nonpsychopaths.

It was one of those rare times in academic life that a research plan came together perfectly. Psychopaths were showing deficits in exactly the regions we had predicted. Their brains are abnormal. I was staring at the first direct scientific evidence of how their brains were different from the brains of the rest of us.

Tears unexpectedly welled up in my eyes as I looked at the results.

I printed off the figures and ran down the corridor to see if any members of the team were in their offices so I could show them.

Dr Liddle was in his office on the phone. I paced back and forth in front of his door as he finished the call. When he hung up, I walked in and handed him the results.

He looked at them, smiled, and said without hesitation: 'We have to send this to *Science*.'

'Do you think *Science* would publish something on psychopaths?' I asked.

'Yes. This is too exciting. The editors might not even know what to do with it,' he replied. 'Make this your top priority; let's get this submitted. And go show the rest of the team. It's remarkable.'

Science magazine is considered the top science journal in the world. The editors publish only a small fraction of the papers sub-mitted to them. For most academics, having a paper published in *Science* is a highly coveted prize.

I returned to the lab and started writing up the results of the first fMRI study of criminal psychopaths. I agonized over the text, wanting to make sure I found the right balance and tone for the readers of *Science*.

We submitted the manuscript a couple of weeks later and anx-iously awaited the e-mail telling us our fate. A week passed with no word. I went back to prison to keep myself distracted from thinking about our paper.

The editors of *Science* wrote back ten days later that they had decided to send our manuscript out for peer review. Some 90 percent of papers don't make it past the first round of editorial scrutiny at *Science*. We had passed the first hurdle.

Another two weeks crept by as we waited on the peer review. I had been incessantly checking my e-mail every few minutes. I nearly jumped out of my seat when the e-mail from *Science* arrived. The scientists who reviewed our paper loved it — my heart leaped. The reviewers had only minor comments that they wanted us to ad-dress. The editors of *Science* invited us to send the revision back to them. The team was ecstatic. I did my best to answer all the review-ers' comments and I sent it back in.

Another two weeks passed; then came an e-mail that I would never forget. Our paper had ultimately been rejected. The editors wrote that while they loved our results, they worried that the sam-ple size was too small and the conclusions society might draw from our research were too dramatic. Our work, they said, had enormous implications for the legal system, and before publishing in *Science*, the editors wanted to make sure the results could be replicated in a larger sample of psychopaths.

I was in shock. We had spent years working on this research. Most of the team had literally risked their lives to complete the study.

But I realized that the editors of *Science* were correct. Our sample size was small. We had only eight psychopaths and eight nonpsy-chopaths in our study. It had been too difficult to transport more of them. Our sample size was fairly typical for early MRI studies of psychiatric populations, but it did raise some important scientific issues that a larger sample size would resolve.

Our team convened to discuss options. We decided to send the manuscript to a top psychiatric journal, *Biological Psychiatry*. The peer review came back quickly, and our manuscript was accepted to be published. The first ever fMRI paper on criminal psychopaths was complete. The team celebrated, but I did not share their enthusiasm. I was frustrated; I wanted more. I wanted to scan more brains of psychopaths, and I never wanted to be criticized again for not collecting enough data. I had to find a solution.

The Path Forward

With our first fMRI study behind us, I returned to my brain wave studies and interviewing more inmates. In my spare time I wrote up a series of five studies for my PhD thesis and submission to peer-review journals.

As my PhD thesis was nearing completion, I planned to jump through the final hoop, the University Defence. At UBC you defend your PhD thesis orally, and the department invites the entire university community to participate in the defence.

Other graduate students had recommended publishing all my PhD research studies in peer-reviewed journals prior to my University Defence. In this way you can always say to your examining committee that your work has met the metric for which all scientists are judged – publication in peer-review journals. If you publish your work prior to your PhD thesis, the thinking, at least of the graduate students, is that you will sail through your final defence and be crowned with your PhD in psychology.

So I wrote up all five of my studies and published them before my final University Defence. In fact, I had collected so much data at the prison that I wrote up another fifteen or so papers and published them too. I actually should have written up a lot more of the data I collected; the data still sit in binders on the bookshelves of my office – a reminder that I have to publish or perish.

While I waited for my PhD thesis committee to set the final date for my University Defence, I went back to prison to keep my mind occupied.

I'd been asked by Dr Brink to conduct risk assessments on inmates. In Canada, risk assessments were done on all inmates prior to release. These assessments included interviewing the inmates and reviewing their files. The information collected from these sources was then used to complete a formula. The formula weights the risk factors and the mitigating factors to determine a score. A low score meant the inmate was low risk to commit new crimes; a high score meant that the inmate was likely to reoffend within three years. The scores are typically used by parole boards to set restrictions on the inmates as they are released.

For example, if the inmate receives a high-risk score, the parole board may recommend that he starts with day parole rather than full parole. Day parole would mean that the inmate was in a secure facility at night but was allowed to go out during the day. In this way the inmate was slowly reintegrated into society, and he was gradually allowed more privileges. These management plans essentially tried to minimize the risk variables and increase the mediating variables to promote a safer society. In my opinion it's good practice for everyone, including and especially the inmate.

When I conducted risk assessments, I would always ask the inmates if they wanted to return to prison. 'No' was the ubiquitous answer. I would tell them that this interview was designed to work out how to reduce the chances that they will get in trouble again after they are released. Many of the inmates enjoyed learning about what variables promote risk and what variables help avoid the risk to reoffend.

Although it was not originally designed for this purpose, scores on the Psychopathy Checklist uncannily predict which inmates will commit new crimes and which inmates won't. Indeed, inmates who score high on the Psychopathy Checklist are four to eight times as likely to recidivate than inmates with low scores. Inmates who get low Psychopathy Checklist scores love the test; inmates who get high scores don't like it so much.

One day while I was interviewing an inmate, another inmate came down and started knocking loudly on my door. Through the window in the door we could see who was pounding on my door. The inmate in my office volunteered to come back later and finish up the interview. He didn't want to get in the way of the inmate at the door. I agreed.

The inmate at the door was a very high-scoring psychopath named 'Martin'. Martin got in a lot of trouble at the prison, and he had a bad reputation. I opened the door to let the other inmate out and let Martin in.

Martin paced around the office, clearly agitated, and then sat down and tossed a piece of paper onto my desk. I retreated to the other side of the desk, picked up the piece of paper, and sat down.

'What can I do for you, Martin?' I asked.

'You can tell me what the fuck that Psychopathy Checklist score means? I just got my risk assessment done and the doc told me I was a psychopath. He said it meant that I was very high risk to reoffend. I'm no Hannibal Lecter,' he said.

I looked down at the photocopy of the Psychopathy Checklist score sheet. Martin had scored a 37. I had scored Martin a bit higher than that during his research interview − a 39 − but when you are in the 99th percentile, as scores of 37 and 39 out of 40 represent, it doesn't make too much of a difference that I had scored him in the 99.8th percentile versus the 99.5th.

'I can tell you about the Psychopathy Checklist. We use this test and many others in our research.'

'Good. What the fuck is this stuff about, this *Lacks Empathy*, *Lacks Remorse*, shit.' He was so angry he was spitting saliva as he spoke.

'You remember when you told me about your victims?'

'Yes,' he replied.

'You told me that you would do it again?' Martin had raped several women very brutally.

'Sure. Whatever. Those bitches deserved it. What's your point?'

I paused, waiting to see if Martin would clue in to the lack of empathy expressed in his last sentence. He didn't.

'Well, your attitude and inability to understand the impact of what you did to those women contributes to your score on *Lacks Empathy and Remorse*.'

'Oh. Well, that sucks,' was his reply.

'Remember when you told me that you screwed your boss out of that deal and when you extorted money from your parents?'

The light was beginning to shine for Martin.

'Fuck. That's all that score means?' he said.

'Yes,' I answered honestly.

'And these other items, what do they mean?' he asked.

I explained some of the other items to Martin.

He just nodded. The Psychopathy Checklist items fit him in all respects. In all aspects of his life, Martin was a classic psychopath.

'Christ. That's all those scores mean,' he said, smiling now.

'Yes, that's it,' I replied again.

'Well, this psychopath thing really sucks; I don't want to be called a psychopath.'

I just looked back at him with my best poker face.

Then Martin scooted forward to the end of his chair, grabbed a pen off my desk, and turned his photocopy of the Psychopathy Checklist score sheet around and started to scribble. He crossed out the word *Psychopathy* from the top of the page and then wrote in big block letters SUPERMAN. He turned to show me his creative work and then said: 'I'm no psychopath. That's the wrong term for me. I'm renaming this the *Superman Checklist*. And now I'm Superman.'

My poker face broke with a smile. I might have to elevate Martin's score on Psychopathy Checklist Item 1 — *Glibness and Superficial Charm*.

Martin stormed out of my office and for the next several days he showed everyone his score on the *Superman Checklist*. Martin told his cellmate that he had to call him Superman or he would beat him up.

Prison is never boring.

·   ·   ·

The day of my University Defence arrived. I dressed in my only suit, and I went over my slides one more time just to make sure my presentation was ready. The auditorium was packed with about two hundred members of the scientific community. Apparently, a thesis defense on psychopaths was interesting to a large number of people.

I gave a thirty-minute presentation and gratefully received the audience's applause before the questions began. I took my time answering questions, and I kept my answers brief. My defence lasted only about an hour. My committee then asked me to step outside while they pondered the pass/fail verdict.

The door opened a few minutes later, and my graduate school mentor, Dr Robert Hare, extended his hand to me and said, '*Dr* Kiehl, would you like to come back in the room?' I'd passed!

I now possessed a PhD in psychology and neuroscience from the University of British Columbia. The crowd applauded as I returned to the room and I thanked them for coming.

My committee nominated, and the university president agreed, that my PhD thesis be noted for 'Distinction', the highest university honor, bestowed upon only a select few theses.

I returned to Dr Liddle's laboratory as a *postdoctoral fellow*.

Not long after I received my degree, Dr Hare retired from the university but he remained a very active researcher. I would be his only student to publish brain wave and brain imaging studies on psychopaths.

Just three months into my postdoctoral fellowship with Dr Liddle, he called me into his office for an urgent meeting. He explained that for family reasons he and his wife were moving their family back to England. I was a bit shocked. I had signed on to work with him for three years.

I realized that I had to start looking for a new job. And I also knew that my time working in Canadian maximum security was coming to a close.

Crash

I returned to my comfort zone – priso – to continue working on brain wave studies using a new EEG system we had just built. This EEG system had sixty-four channels, more than eight times my previous system. The extra channels allowed me to cover the entire head during a recording session rather than the sparse sampling of the eight-channel system I had started off with.

I also updated my curriculum vitae and sent it out to all the academic jobs around the world that might fit my background. To keep myself from worrying about where life would take me, I kept my head buried in my prison research. I had returned to my schedule of arriving at the prison before dawn and working through a lunch of peanut butter and jelly sandwiches.

One dark, rainy Wednesday, I left prison at 5 p.m. so that I could head to the university MRI to work with Dr Mackay and his group for one of our evening research sessions. His PhD students had asked if I would let them scan my brain that night.

The windscreen wipers on my little Toyota truck had worn out, and I was having trouble seeing the highway through the pouring rain. I slowed down to try to improve visibility. The intermittent shine of the streetlights spaced far apart made it unusually difficult to see the road where there was no light. I slowed down some more.

A loud *whoosh* exploded by my left ear, and my startle response nearly put me through the roof of my Toyota. A wave of water engulfed my pickup and I swerved to avoid the onslaught. My pickup fishtailed and I instinctively countersteered, hoping that I would not hydroplane off the road. As the wipers worked furiously, I was able to get the vehicle pointed again in a straight line. I realized that I had been passed on the left by a large truck. As the truck had passed, we had both gone through a flooded section of the motorway.

My headlights must have danced all over the place, indicating to the trucker that I was in trouble. The trucker put on his flashing hazard lights and slowed down, and then signalled he was pulling over at a rest stop.

My hands were squeezing the steering wheel so tightly that I could barely feel them. I instinctively followed the trucker down the entrance to the rest stop. Adrenaline was coursing through my body.

I pulled in next to him in the car park. He jumped out of his cab and walked over to me. As I rolled down my window, the rain started to pour inside.

'You okay?' he asked.

'Yes, I think so,' I managed to reply.

'You really fishtailed all over the place; that was a great recovery.'

'Thanks.' I didn't bother telling him that it had been sheer luck that I had not headed off the side of the motorway and tumbled down the mountain.

'I'm really sorry. I didn't see the water on the road. It nearly caused me to lose control too.'

'It's all okay,' I said. 'Everything is okay. Nobody got hurt.' I looked over at his truck for the first time to see if there was some damage from the wave of water. The words *Magnetic Resonance Imaging* were written in blue ink on the side of the truck's trailer. I read the signage again.

'What's in your truck?' I asked.

'It's an MRI.' came the reply.

I pushed my door open, rudely bumping the driver.

'Hey, what are you doing?' he said as he recoiled back against the side of his truck. 'It was an accident,' he stammered. 'I don't want any trouble.'

'No, no,' I said as I held up my hands in a show of peace. 'I'm not angry. I want to know about what's in your trailer.'

Clearly relieved, but still unclear what I was doing, the driver asked, 'Why do you care what's in my trailer?'

I told the driver, 'I study criminal psychopaths. Does your trailer have an MRI that can be used for research? If so, can I take it into a prison?'

The driver backed away from me; the fear on his face was clear. 'You wanna do what?' he asked.

I put my hands up again in a placating gesture. 'Sorry,' I said. 'Let me start over.'

'I'm *Dr* Kiehl. I'm a forensic psychologist who studies psychopaths. I'm trying to figure out what's wrong with them.'

I took a step back to take in the size of the truck.

It was a full-size semi trailer, with two axles at the back that were spread a bit farther apart than normal. There was a large belly compartment that sat just inches above the ground. I figured that belly compartment must have contributed to amplifying the wave of water that had engulfed my pickup. There was a door about four feet off the ground near the middle of the trailer.

'Can I take a look inside?' I asked.

'Sorry,' he replied, 'I'm not supposed to open the doors until I get to the next location.'

'Come on,' I said, 'you almost ran me off the road. The least you can do is let me take a peek inside.'

'Okay, but just for a second. I have to get this to the site in Surrey.' Surrey was a suburb outside of Vancouver.

The driver pushed a button on the door of the belly compartment and it popped open. He flipped a switch, and I heard the locks snap open.

He walked over to a small hatch under the door and pushed on it. It popped down, then he reached in and pulled out a retractable staircase.

He went up the stairs and pulled a key from his pocket and unlocked the door, pulling it open. I followed him inside.

He flicked four light switches on, and the room was bathed in a bright light. I squinted to allow my eyes a second to adjust. The trailer consisted of three rooms. To my right was a door that contained all the hardware that ran the MRI. The middle room, where I was standing, was a long, narrow control room with a desk and chair. To my left was a door with a window next to it. I could see the MRI at the far rear of the trailer – positioned right in the middle of those two oddly spaced axles, balancing the weights, I assumed.

I was drawn to the door to the MRI room, as if I was being pulled in by the magnetic field.

'Don't go in there,' the driver said. 'The MRI is still on. I don't know why, but they told me they never shut it off.'

I started to explain the physics of the superconductive magnet sitting in that room, but then stopped and simply told him that you can't shut them off; it costs too much to ramp them back up again.

'Oh,' he replied. But his tone indicated that my knowledge of the machine's engineering had convinced him I was who I said I was.

The driver explained that the mobile MRI was used for clinical patients in the lower mainland of British Columbia. It travelled to where there were populations of people who needed scans, but where there were no MRIs available in the local hospitals. The driver's full-time job was to drive the machine around.

The mobile MRI was not the type or quality that we needed for our research, but it planted a seed in my brain. I knew that as the technology advanced, engineers would figure out a way to put the best MRI scanners in a trailer. And then I might be able to take one into prison and scan lots of psychopaths.

I thanked the driver and sat there in my Toyota as he pulled away. *Someday*, I thought.

chapter 6

Bad Beginnings

ғacт: ᴀ psychopath is born every 47 seconds.[1]

Parents of children with severe disruptive behavioural problems often write me letters asking for help. The parents frequently start out with a frank declaration of their trepidation about sending me a letter because it means they have arrived at the dark realization that their child could be a budding psychopath. No parent wants to confront, much less accept, that fact. I have a rather large file where I keep copies of these letters. Occasionally, when I run into a stumbling block or have a difficult time doing research, I'll return to the parents' letter file and read a few. It gives me an extra surge of motivation to hurdle the stumbling block or push through the difficult period and keep focused on the prize — developing an adequate answer to the questions parents raise in their letters.

The letters are heart wrenching to read; I can't imagine how difficult they are to write. They come from around the world, but the letters share a common theme. Parents typically start by saying they have been dealing with their child's problems for many years. From the beginning, they struggled to understand why their child had difficulties developing attachments to them, particularly to the mother. The troubled child was impulsive and angered easily, much more easily than other siblings or children with whom the parents had experience. Fights were common and often left the other party injured and their child unwilling, or unable, to accept responsibility or show guilt. Sanctions or punishment did not change the pattern of impulsivity or anger. Parents often raise, with difficulty, incidents of animal cruelty. Goldfish are often found outside the bowl or trapped someplace where they can't swim, like at the bottom of a sink where the water has been drained out and there is just enough left for the fish to flop around in. Hamsters, gerbils, and other such animals mysteriously disappear or are found dead in their cages. Abuse often escalates to injury to or death of cats and dogs.

The child usually had significant and repetitive problems at day-care, preschool, and nursery; and as school progressed, the repertoire of behavioural problems expanded: repeatedly hitting other children with foreign objects, throwing rocks, and getting into playground fights, all of which resulted in visits to the head teacher's office.

The child has little interest in developing friends or spending time with them. The troubled child seems to prefer to be on his own (these children are usually male). Affect is limited or lacking in every aspect of the child's life. Attachments to people are weak or feeble, and such children show evidence of a profound inability to understand relationships and social behaviours such as sharing or co-operation. Some parents think their child has an undiscovered emotional learning disorder; they want to know if I agree with that assessment.

One of the most telling aspects of the letters is that none of the children are described as normal from birth. Parents say they noticed something different, odd, or abnormal about the child from the very beginning. The letters never describe a normal child, followed by a sharp transition in behaviour precipitated by something like a head injury or stressful event (i.e., divorce or death of a parent or sibling). They never describe a period of time where the child went from a normal state into a slow decline. Such a pattern is seen in a number of mental illnesses and is commonly referred to as a *prodromal period* – a slow, progressive change in normal functioning until a real mental break occurs. These children's histories are presented as qualitatively different from other siblings or children from a very early age; if anything, the behaviour has been disappointingly consistent, if not worsening, since birth.

I am never the first mental health professional that parents have sought advice from. Parents usually start asking questions in the doctor's office. It is not uncommon for the parent to say that the child was first diagnosed with attention-deficit hyperactivity disorder (ADHD). Following administration of amphetamine (the main ingredient in medications for ADHD such as Ritalin or Adderall), the child's behaviour initially slowed down but then exploded with an even more bewildering array of behavioural problems once the child's system adjusted to the stimulant effects. A return to the doctor's office will often lead to another diagnosis, then another, and an associated cocktail of medications. Childhood bipolar disorder and autism spectrum disorder are diagnoses du jour that are often used to (mis)label the child with severe disruptive behavioural problems. Frustrated, parents turn to school psychologists, then to professional clinical psychologists, to a second and then a third opinion from child psychiatrists. And then some parents turn to academics. The parents look to academics like me because they feel the community mental health providers are missing something. The parents have educated themselves about the various diagnoses their child has been given, and they often disagree with the child's doctors. They have cataloged and reviewed the symptoms their child has, and then they realize some of the symptoms fit the definition of *psychopathy*. Sometimes they feel like they have discovered a new disorder that needs to be studied. Most to the point, they just want to know what they can do to help fix their child. The stress is palpable in every letter.

A number of my colleagues who study psychopaths have told me they too are often contacted by parents with troubled children looking for help. I meet up with my colleagues every year at annual scientific conferences. There is usually at least one conversation at dinner about a particularly troubling letter from a parent or relative looking for help. We ask one another whether there are any new treatments, drugs, help groups, or other options that have been developed in the last year to help answer the parents' questions and address their pain.

At these dinners, we also often discuss clinical cases of adult psychopaths that have come across our desks in the past year that were illustrative of the condition. Someone will often note the similarity between the child described in the parents' letters and the case history of the latest psychopath who showed up for a prison research study.

Much can be learned from examining specific case studies of psychopaths. Two cases that I will review in great detail are those of 'Brian' and 'Eric'. In this chapter, I will present the childhood of these two boys. Later, I will examine their adolescent and adult years to demonstrate the symptoms of psychopathy across different developmental periods of life.

Brian

Brian was born in the idyllic town of Nashua, New Hampshire, located in the southeast corner of the state, about an hour's drive from the closest major city, Boston, Massachusetts. Nashua has been named *Money* magazine's 'Best Place to Live in America' and is the only city in the country to ever win the honour twice. Brian is the second child born to parents Genevieve and James; he is two years younger than his sister, Hillary. The family would eventually grow to five children, with the additions of Stephen, Jeffrey, and Jimmy.

His mother went into labour prematurely with Brian, and prior to the attending physician's arrival, an emergency room nurse reportedly pushed Brian's crowning head back into the birth canal and forcibly crossed his mother's legs to prevent delivery. Brian was born an hour later.

Brian suffered from migraine headaches throughout childhood. Doctors are unclear whether the migraines are related to the birth complications, but it seems not unlikely that the traumatic birth played some role. The headaches are amplified by stress or excitement and are so severe they lead to vomiting. As an infant, Brian banged his head against the baby cot and the walls of the house until he was black and blue. These problems continued until he was three and a half years old, and headaches would plague Brian through his teenage years.

Brian's father, James, was a travelling salesman. Shortly after

Brian's birth, James was back on the road, and he was absent for the majority of Brian's and his siblings' childhoods. Largely raised by their mother, the five children lived in a typical home for the area, just a few miles outside of town. Genevieve adopted an authoritarian parenting style trying to manage the five children. Punishment was strict and discipline was harsh. A belt, sometimes with the buckle end, was used to punish the children. Brian's brothers and sisters reported that their mother often hit the children so severely she would wear herself out. These beatings often took place in front of the other children so that they would learn from observing the punishment. The beatings were unpredictable and severe.

Both parents drank heavily and were reported to be 'functioning alcoholics'. However, before Brian started nursery, James's drinking became worse and contributed to his losing his job. Unable to pay their bills, the family was evicted from their home and forced to rent, with the children crowded into shared bedrooms.

Brian was a chronic bed-wetter, a problem that infuriated his mother. She would make Brian sleep in the soiled sheets. Childhood acquaintances would report the home smelled like urine.

At the age of seven, Brian and his five-year-old brother, Stephen, set fire to the detached garage. Even after being harshly disciplined, the two brothers continued to play with matches. They would throw lit matches at each other, daring the other to see who could sit still the longest as matches flew by. Their mother once caught them and held their hands until the matches burned down and seared their palms. In addition to bed-wetting and fire-setting, Brian had other behavioural problems in childhood. Also when he was seven, Brian chased his older sister around the house with a large kitchen knife. When his sister evaded him and locked him in the basement, Brian stabbed through the basement door in anger.

Brian didn't make many friends growing up. His behaviour was risky and impulsive, and he had accidents and got into fights. One fight with his brother Stephen resulted in his front teeth being knocked out.

From an early age, Brian was cruel to animals. When he set fire to a snake, the resulting blaze ended up spreading to a field and firefighters had to be called in to save the neighbourhood from going up in flames. He poured petrol on a cat and lit the animal on fire, burning it to death. Brian also reportedly abused the family dog.

Brian's schoolwork performance was sloppy, and his grades were average to below average, even though he had no reported learning disabilities. Brian's attention span was limited, and he was unengaged by school or after-school activities. He did not participate for any significant period of time in any group sports. He began smoking cigarettes by age eleven.

When Brian was in fourth grade, the family moved to Lisle, Illinois. His mother worked as a playground supervisor at a local school, while his father took a job as a travelling salesman at a new company and was again away frequently. By age thirteen, Brian was doing drugs and engaging in burglaries. He had sex for the first time with an eleven-year-old girl. He had multiple contacts with the police and childhood charges that included disorderly conduct, vandalism, and burglary.

Before his fourteenth birthday, Brian had had sex with multiple partners, including a twenty-year-old married woman. His drug use escalated from nicotine and marijuana to PCP, LSD, methamphetamines, cocaine, Quaaludes, hashish, codeine, Valium, inhalants, and alcohol. Brian was removed from his home by social services and placed in a boys' home. He promptly ran away. Before his fourteenth birthday, he landed in detention for a series of burglaries.

Eric

Eric was from the south side, the tough side, of Milwaukee, Wisconsin. He grew up in a very unstable home not unlike many of the homes in the neighbourhood. Exactly who was taking care of him when he was very young is difficult to determine. However, when he started school at age seven, he was living with his mother, with periodic visits from his father, who sold heroin, cocaine, and marijuana. His mother was addicted to cocaine. It was not clear if his parents ever lived together for any length of time, but both had multiple sexual partners who lived with them at times. Eric witnessed several fistfights between his parents. His father was arrested during one fight and charged with domestic violence.

About the time he started nursery, Eric's mother had been arrested for dealing cocaine. His father was also in prison at the time, so Eric was sent out of state to live with an aunt and uncle. He claimed to have no emotional attachments at this time in his life. Later, Eric reported that he had got into a lot of fights in primary school and never made any friends. He described himself as an outsider. He said he preferred it that way, and he liked being different from the other children.

Eric developed a taste for drugs and alcohol from his father and his father's girlfriends. His father began to supply him with alcohol when he was in first grade, although Eric did not like the taste or effect it had on him at first. By age eleven, he was back living with his father, who had just been released from prison. His father supplied him with marijuana and cocaine and encouraged Eric to begin to traffic in drugs. Predictably, within two years, his father was back in prison, this time on multiple counts of armed robbery. Meanwhile, Eric's mother was undergoing detox in a women's prison. Eric's aunt and uncle were unwilling to take him back because he was too disruptive, so he was sent to live with his grandmother in another state.

Eric spent little time actually living in his grandmother's house. Instead, he roamed the streets, travelling to other parts of the country and living 'by my wits'. He was twelve years old. He used several fake IDs to get around. He earned spending money by selling drugs, 'hustling dice', selling things he had stolen, and 'conning' other kids out of money or possessions with rigged gambling swindles. He was adjudicated delinquent on charges of trafficking marijuana. He had sex for the first time while on the road, a one-night stand with a younger girl. He had about twenty sexual partners during this time, all casual acquaintances. Many of the girls he slept with gave him money or a place to stay. He would steal from, con, or manipulate the girls and their families until they would kick him out or until he moved on.

At the age of thirteen, when his mother was released from prison, he was returned to her custody. Over the next nine months, Eric was charged with twelve more offences and was placed in a series of foster homes, community facilities, and eventually in corrections.

Budding Psychopaths?

Every adult psychopath that I have worked with was different from normal children from a very early age. Their prison files are typically replete with stories from siblings, parents, teachers, and other guardians about how as a child the psychopath was emotionally disengaged from other siblings, got in trouble more frequently, engaged in more severe antisocial behaviours, and started using alcohol and drugs and having sex at a younger age than other children. Most adult psychopaths report that as children they didn't have close friends, didn't feel the need to participate in any group activities like football, and if they did participate, they tried to cheat, got in fights, and generally did not enjoy playing with others. Psychopaths typically don't get along well with their parents and often are very distant from their siblings. The nature of their early childhood often results in them being labelled the black sheep of the family.

Many adult psychopaths have grown up in home environments not unlike Brian's and Eric's. It is perhaps not surprising that the two of them ended up in juvenile detention. Indeed, it may seem as if both were destined for a life of crime based on their early family environments. However, I should note that just as there are many individuals who come from terrible environments who end up in prison, there are many more who are resilient to such upbringing and environments who do not end up in prison as youth or adults. In fact, only a fraction of children who come from such terrible environments end up becoming criminals. Moreover, psychopaths often come from stable middle-class and even upper-class families. The disorder does not discriminate – psychopaths are found across all socioeconomic strata.

For example, 'Brendan' is an example of an adult psychopath who came from an above-average home environment. Brendan's parents were both doctors. He had been raised in an affluent, gated community, and he attended private schools from childhood. He had two 'unaffected' siblings; that is, his older brother and younger sister did not have any affective symptoms or behavioural problems. Indeed, his older brother was attending Harvard University when Brendan committed his index offence. During our interview, Brendan kept noting that he did not belong in prison with the other inmates. He viewed other inmates with disgust, and he had a very elitist attitude.

Like all adult psychopaths, Brendan had been different from his peers and siblings from a very early age. Brendan told me that as a child he would create vicious games where his dog had to run a gauntlet of traps in order to be fed. Brendan's parents recognized his early behavioural problems and had sent him to see professional psychologists and psychiatrists. Brendan had been disciplined at the private schools for fights, cheating, setting fire to the gym, theft of lunch money, and other crimes. His parents' wealth had kept him from developing serious criminal convictions as a juvenile. Indeed, as a teen he had stolen a sports car from the garage of one of the other students' parents; Brendan had wrecked the car and quite miraculously had not been seriously injured. Brendan's parents paid for a new sports car, and the theft was kept out of the juvenile courts. However, Brendan's parents could not keep him out of jail for cold-blooded murder.

As a young adult Brendan had decided to kill his ex-girlfriend's new boyfriend. He tried to make it look like it was just a fight that escalated out of control, but the police uncovered details that indicated Brendan had been planning the event for some time. His parents hired the very best defence lawyer money could buy, and Brendan received a very good plea deal. He was going to serve only about five years for manslaughter, and he had managed to be sentenced to a minimum-security facility. The repertoire of behavioural problems from an early age, escalating into more severe antisocial behaviour as a teen, and finally culminating in a severe crime as a young adult is a very common trajectory for psychopaths. Brendan was no different from the hundreds of other psychopaths I have studied.

Regardless of the child's early environment, it is the severity of the psychopaths' childhood misbehaviour that stands out in contrast to their siblings and peers.

. . .

What are some of the characteristics that Brian and Eric displayed that might help identify them as different from other children, even other children from similar environments? Can we predict whether Brian or Eric will develop into psychopaths? That is, can we figure out which (pre)psychopathic symptoms in childhood can be assessed accurately and reliably, and, most important, does the assessment of these symptoms in childhood predict which youth will be psychopaths as adults?

Through the stories of Brian and Eric we see numerous behaviours that might suggest disturbances consistent with developing into a psychopath as an adult. Brian set a number of fires and abused animals. He also wet his bed through childhood. These three symptoms are known as the *MacDonald Triad*, and decades ago they were thought to be precursors to future homicidal behaviour, even possibly serial homicidal behaviour.[2] However, subsequent research has failed to show any statistically significant predictive relationship between the MacDonald Triad and future homicidal behaviour. Yes, many people who commit homicide as adults had these symptoms as children, but the vast majority of children who have these symptoms do not go on to commit homicide as adults.

Since the 1960s there has been a lot of research on the causes of bed-wetting, or enuresis. It turns out that a number of problems can lead to enuresis.

First, most children who suffer from this—about 85 percent— grow out of enuresis by age five. However, a small percentage of children continue bed-wetting until age ten or later. Researchers have determined that involuntary control over the bladder occurs as a result of at least four different neuronal paths.[3] Developmental abnormalities or delays in any one or more of these neural circuits can lead to chronic bed-wetting. One of the four paths leads through the region of the brain known as the amygdala. As I mentioned in Chapter 5, the amygdala acts like an amplifier in the brain and helps push into awareness any salient stimulus. The amygdala amplifies some stimuli that we encounter automatically, like angry faces or snakes. It can also amplify things we learn are important, like emotional words or darkly clad individuals.

My hypothesis[4] is that it's the amygdala bladder circuit that is abnormal in youth who go on to commit homicide as adults. If that is the case, we need to revise the MacDonald Triad to indicate that the risk for future violence is present in chronic bed-wetting youth only if the circuit responsible is the one that passes through the amygdala, something that can be tested using modern neuroscience techniques. It's possible the revised MacDonald Triad could improve the likelihood of the presence of the three symptoms (bed-wetting caused by failure in the amygdala circuit, fire-setting, and animal cruelty) to predict a child who is on a trajectory towards future severe antisocial behaviour as a teenager and adult.

In Brian we also see evidence of violence and aggression in multiple domains of his life: severe animal abuse, even torture. We see violence towards his sister and brothers. Brian's antisocial behaviour started from a very early age. Moreover, he commits burglaries and other crimes alone. He doesn't need a peer group, a potent influence in adolescent years, as an impetus for his antisocial behaviour. We see little evidence of his antisocial behaviour diminishing with punishment or brief periods of incarceration. We see precocious sexual behaviour early on; Brian is promiscuous, and all his relationships are brief. Brian seems disinhibited in many domains of his life. His emotional detachment – the absence of close friends, lack of participation in group activities like football, difficulties with his siblings and parents, interpersonal aggression – all suggest that Brian is on a trajectory towards psychopathy.

Eric shows a lot of the same problems as Brian. He gets in a lot of fights at school and doesn't participate in group sports. Eric is a drifter and he acts alone in most of his antisocial behaviour. A hustler, he uses schemes to earn money while on the run. He also begins sexual activity from a very early age and is highly promiscuous. He seems to have problems maintaining long-term relationships. He engages in numerous, serious criminal activities starting very early in life. However, details about Eric's background are sketchy, as there were limited sources of information about this time in his life. Whereas he readily admits to his criminal activity, we have no information on whether he engaged in other risky behaviours like animal cruelty or physical assaults on other people.

Brian and Eric have shown us a lot of reasons to think they are on a trajectory towards lifelong antisocial behaviour, and they appear to exemplify a number of psychopathic traits. Their childhood behavioural histories are similar to those of hundreds of adult psychopaths I have interviewed. But do they meet the criteria for psychiatric diagnoses as youth?

Childhood Diagnosis of Conduct Disorder

Personality disorders are by definition an enduring pattern of thinking, feeling, and behaving that is relatively stable over time. Indeed, in children and adolescents, symptoms of personality problems must be present for a significant period of time (typically more than six months), and not just a reaction to the social environment. The American Psychiatric Association's *Diagnostic and Statistical Manual of Mental Disorders (DSM-IV-TR)* uses the terms *Conduct Disorder* and *Oppositional Defiant Disorder* to describe youth who have significant disruptive behavioural problems. The *DSM-IV-TR* symptoms of conduct disorder and oppositional defiant disorder are listed in Box 3:

BOX 3

Conduct disorder and oppositional defiant disorder are diagnosed based on the following criteria.

CONDUCT DISORDER

A. A repetitive and persistent pattern of behaviour in which the basic rights of others or major age-appropriate societal norms or rules are violated, as manifested by the presence of three (or more) of the following criteria in the past 12 months, with at least one criterion present in the past 6 months:

Aggression to people and animals
   (1) often bullies, threatens, or intimidates others
   (2) often initiates physical fights

(3) has used a weapon that can cause serious physical harm to others (e.g., a bat, brick, broken bottle, knife, gun)

(4) has been physically cruel to people

(5) has been physically cruel to animals

(6) has stolen while confronting a victim (e.g., mugging, purse snatching, extortion, armed robbery)

(7) has forced someone into sexual activity

Destruction of property

(8) has deliberately engaged in fire-setting with the intention of causing serious damage

(9) has deliberately destroyed others' property (other than by fire-setting)

Deceitfulness or theft

(10) has broken into someone else's house, building, or car

(11) often lies to obtain goods or favours or to avoid obligations (i.e., 'cons' others)

(12) has stolen items of nontrivial value without confronting a victim (e.g., shoplifting, but without breaking and entering; forgery)

Serious violations of rules

(13) often stays out at night despite parental prohibitions, beginning before age 13 years

(14) has run away from home overnight at least twice while living in parental or parental surrogate home (or once without returning for a lengthy period)

(15) is often truant from school, beginning before age 13 years

B. The disturbance in behaviour causes clinically significant impairment in social, academic, or occupational functioning.

C. If the individual is age 18 years or older, criteria are not met for Antisocial Personality Disorder.

Code type based on age at onset

312.81 Conduct Disorder, Childhood-Onset Type: onset of at least one criterion characteristic of Conduct Disorder prior to age 10 years

312.82 Conduct Disorder, Adolescent-Onset Type: absence of any criteria
characteristic of Conduct Disorder prior to age 10 years

312.89 Conduct Disorder, Unspecified Onset: age at onset is not known

Specify severity

Mild: few if any conduct problems in excess of those required to make
the diagnosis and conduct problems cause only minor harm to others
(e.g., lying, truancy, staying out after dark without permission)

Moderate: number of conduct problems and effect on others intermedi-
ate between 'mild' and 'severe' (e.g., stealing without confronting a
victim, vandalism)

Severe: many conduct problems in excess of those required to make the
diagnosis or conduct problems cause considerable harm to others
(e.g., rape, physical cruelty, use of a weapon, stealing while confronting a victim, breaking and entering)

OPPOSITIONAL DEFIANT DISORDER

A. A pattern of negativistic, hostile, and defiant behaviour lasting at least 6
months, during which four (or more) of the following are present:

(1) often loses temper

(2) often argues with adults

(3) often actively defies or refuses to comply with adults' requests or rules

(4) often deliberately annoys people

(5) often blames others for his or her mistakes or misbehaviour

(6) is often touchy or easily annoyed by others

(7) is often angry and resentful

(8) is often spiteful or vindictive

*Note: Consider a criterion met only if the behaviour occurs more frequently
than is typically observed in individuals of comparable age and developmental level.*

B. The disturbance in behaviour causes clinically significant impairment in social, academic, or occupational functioning.

C. The behaviours do not occur exclusively during the course of a Psychotic
or Mood Disorder.

D. Criteria are not met for Conduct Disorder, and, if the individual is age 18 years or older, criteria are not met for Antisocial Personality Disorder.

DSM-III was the first edition of the *DSM* to include an operational definition for conduct disorder, and the criteria for conduct disorder have fluctuated substantially ever since. *Conduct disorder* was originally defined as repetitive conduct in which the rights of others are violated, including physical violence against persons or property. *DSM-III* was heavily influenced by behaviourist theory, and as such, environmental influences played a prominent role in describing the two subtypes, the Undersocialized type and the Socialized type. The 'undersocialized type' included children who fail to establish normal degree and quality of affection, empathy, or social and romantic bonds with others. In contrast, the 'socialized type' was able to develop normal attachments with others but still got in a lot of trouble. These groups were based on taking all the symptoms of disruptive children and conducting what is known as *factor analyses*, dividing the symptoms, statistically, into two categories.

DSM-IV dropped the distinction between the undersocialized and socialized subtypes. Now the undersocialized group is known as *early age of onset group.* Poor parental monitoring now plays an important role in assessing conduct disorder. Research shows that youth who have greater interpersonal problems at an early age as well as other psychosocial risk factors (i.e., poor parenting) have more stable antisocial traits into adulthood.

DSM-IV listed four general conduct disorder categories: aggression to people and animals, destruction of property, deceitfulness or theft, and serious violations of rules. For a child or youth to receive a diagnosis, at least three of fifteen symptoms from these headings must be present for at least twelve months.

In reviewing what we know of Brian's and Eric's histories, we can see that both Brian and Eric meet criteria for *severe* conduct disorder, childhood-onset type. And since they meet criteria for this more severe disorder, psychologists would not give them the lesser diagnosis of oppositional defiant disorder (even though they meet most of the criteria).

What does the conduct disorder diagnosis mean? Is it the precursor to a diagnosis of psychopathy as an adult? Well, it's actually not that clear. The diagnosis of conduct disorder is exclusively based upon observable behaviours; it does not assess any of the emotional, interpersonal, and affective traits associated with psychopathy. In fact, there is no mention of lack of empathy, guilt or remorse, or shallow affect in the *DSM-IV* conduct disorder diagnosis. Many scientists have argued that the failure to include such *callous and unemotional* traits in the diagnosis of conduct disorder severely limits its utility. And there are additional criticisms of the conduct disorder diagnosis. Nearly 80 percent of children diagnosed with conduct disorder grow out of it and do not develop an adult personality disorder or psychopathy. This would suggest that conduct disorder is not really a disorder at all. In other words, the diagnosis does not predict which children are on a trajectory towards lifelong personality problems, future antisocial behaviour, or psychopathy.

Perhaps the most poignant criticism of the conduct disorder diagnosis comes from the former president of the American Psychological Association, director of the Yale University Child Conduct Clinic, and author of over seven hundred peer-reviewed manuscripts and forty books, Dr Alan Kazdin. Dr Kazdin has noted that there are 32,647 combinations of symptoms that youth can have to meet the diagnosis of conduct disorder.[5] To the extent that the symptoms of conduct disorder are independent of one another, this means there are over 32,000 different types of children with conduct disorder. It is a clinical psychologist's nightmare. There is no sensitivity or specificity to the diagnosis. It allows for a vast, diverse range of children to be diagnosed with the disorder. And it doesn't predict anything. It's essentially a hodgepodge of symptoms with very little utility.

Within secure juvenile correctional systems where I have conducted research, clinicians often don't even bother doing the assessment for conduct disorder since nearly every youth meets criteria. Conduct disorder simply does not help differentiate children and youth in the criminal justice system. Thus, the diagnosis of conduct disorder suffers from many of the same criticisms as the adult antisocial personality disorder diagnosis (reviewed in Chapter 2).

But this picture is beginning to change. For the last twenty years or so, a number of dedicated scientists have been developing measures to quantify callous and unemotional traits in youth. Psychologists believe that examining both the callous and unemotional traits and the antisocial and impulsive traits in youth will help to identify those at the highest risk for developing into full-blown adult psychopaths.

Callous and Unemotional Traits in Youth

Trying to assess and predict which children will become psychopaths as adults is a difficult task. Some have argued that scientists should not even attempt such an endeavour because one side effect of diagnosing a child as a psychopath is the possibility the label might stigmatize the child. Such labelling could also lead to a self-fulfilling prophecy. Others have suggested that if parents are told their children are psychopathic, it may further the divide between parent and child. Finally, parents of children labelled psychopathic may be stigmatized as well. These are very serious concerns, and my colleagues who study these high-risk children are very sensitive to these issues. Indeed, scientists in the field generally go to great lengths to avoid the term *psychopathic* when discussing these youth. The term that is most commonly used is *callous and unemotional traits*. Among psychologists, youth with significant callous and unemotional (CU) traits and disruptive behaviours, or conduct disorder, are known as *CU/CD*, or callous conduct disordered youth.

It is clear from my own clinical experience, echoed by most other forensic practitioners who work with psychopaths, that nearly all psychopaths were emotionally abnormal as children. So if we want to help understand how the condition develops, and how to answer parents' questions about how to help manage and treat these children, scientists have to try to understand how these emotional symptoms manifest themselves early in life. Indeed, ignoring the problem or not studying the problem is absolutely not the right answer. Careful and thoughtful science is the answer to helping to address psychiatric disorders, and psychopathy in particular.

As I discussed previously, researchers and clinicians who use the

DSM are trained on how to assess and relate the symptoms in the manual to the clients in their care. Typically, a thorough assessment is done on a client based on an interview and a review of the client's life history. It is from this information that trained experts then determine the psychiatric diagnosis according to the guidelines of the *DSM*. However, psychology uses a number of other techniques to quantify and assess psychiatric symptoms.

Given that psychology is the study of behaviour, it is also then the study of the brain, since all behaviour emanates from the brain. Psychologists have devised a number of techniques for measuring and quantifying behaviour.

One method psychologists use to assess personality traits is to give a client a list of questions to answer. These so-called self-report inventories can be quite long. For example, the first version of the popular Minnesota Multiphasic Personality Inventory (MMPI) had 567 questions. Psychologists use the MMPI to assess the various dimensions of personality – like trustworthiness and introversion (the ability to feel comfortable in social settings). Instruments such as the MMPI are useful to develop an understanding of the personality problems the psychologist's client might be experiencing so that an appropriate treatment programme can be developed. Self-report tests are common in psychology, and they have recently been developed to assess CU traits in youth.

The first and most commonly used self-report instrument to assess CU traits in youth is the Child Psychopathy Scale (CPS) developed by Dr Don Lyman of Purdue University. The CPS includes questions asking the child how he (or she) relates to others, what types of things the child considers to be important, how angry the child gets, and so on.

Self-report inventories can be very helpful in assessing behaviour, and they are very popular in psychology. However, self-report inventories have their critics and their limitations. Moreover, some of the limitations of self-report inventories are exacerbated when trying to assess CU traits in young children.

One rather obvious limitation of self-report instruments is that they require the ability to read. Despite dramatic decreases in the rate of illiteracy in the world, it is still a significant problem. And illiteracy is much higher in children who have behavioural problems at school. Thus, many of the high-risk youth whom we want to assess for childhood CU traits may not be able to read and answer the questions on self-report scales like the CPS or the MMPI.[6]

Another limitation of self-report scales in psychology is that they require co-operation. An easy way to defeat a self-report test is to lie, randomly fill in the answers, or just flatly refuse to complete the test. This greatly limits the utility of self-report tests in adversarial contexts.

Finally, perhaps the most important limitation of self-report scales that attempt to assess CU traits is that children with such traits may lack the ability and insight to report accurately on their emotional world. Such a lack of insight may confound researchers' attempts to assess these traits.

To address these issues, researchers have turned to other techniques to assess these traits in children. To complement children filling out questionnaires, researchers now often ask parents and other caregivers to complete questionnaires about their children as well. Of course, some of the same limitations apply to this strategy as they do to self-reports for children, but generally gathering more information is better when trying to assess CU traits.

With parent or caregiver self-reports, researchers try to get both parents to independently answer questions about the behavioural and emotional life of the child. If possible, additional self-report inventories are collected from teachers or other caregivers who have directly observed the child for a significant period of time.

Dr Paul Frick of the University of New Orleans and collaborator Dr Robert Hare of the University of British Columbia have developed a number of parent and teacher scales for the assessment of CU traits in youth, including the Antisocial Process Screening Device (APSD). The APSD is modelled and scored very similarly to the various versions of the Psychopathy Checklist.[7] The APSD is a twenty-item questionnaire that is given to parents or caregivers and a teacher. There is a self-report version for the youth to complete as well. A great deal of research has been conducted with the APSD; it has been used with youth ranging from four to eighteen years of age. In parallel, Dr Frick developed the Inventory of Callous-

Unemotional Traits (ICU), with versions for parents, teachers, and the child to complete. The ICU has versions designed for preschool, nursery, and primary-school-age children; it was developed to address some potential technical limitations of the APSD scales.

When parents are asked about the emotional life of their child, one difficulty that arises is that it can be hard for parents to check off items indicating that their child lacks empathy, is violent, and appears to enjoy inflicting pain on animals. This is especially the case when their child might be involved in the legal system. Sometimes parents will fudge reports about the severity of their child's misbehaviour. Also, since people generally believe that a child's misbehaviour is partially the fault of the parents, parents may be further motivated to limit the reporting of their child's problems. Indeed, as the letters I receive from parents with such children attest, it is a very difficult situation to deal with, and parents often just don't know what to do.

The last kind of test that researchers use to assess CU traits in children (and adults) is an expert-rater device that explicitly assesses these traits, such as the Hare Psychopathy Checklist-Youth Version, which is an age-appropriate version of the adult Psychopathy Checklist.[8] Researchers interview the child, parent, and perhaps other key people in the child's life, and then rate the child on CU traits. The limitations of such 'expert-rater instruments' are that they take a lot of time to complete, are typically more expensive than self-report or parent and teacher reports, and require that the person doing the assessment is well trained. But one advantage of the Youth Psychopathy Checklist is that co-operation by the child, or the parents, is not necessary to complete the test. The Youth Psychopathy Checklist can be completed by reviewing details from teachers, coaches, neighbours, and other caregivers. In this way, it is possible, if necessary, to complete the assessment in the absence of talking to the child or parent, such as when a parole office needs to know how to develop a management or release plan for a young offender.

In my laboratory, my preference is to do all three types of reports on children who have elevated CU traits. We then compare the instruments to see if they generate the same pattern of results and to see if they identify the same children as scoring high or low.

The good news is that there have been considerable advances in the development of these assessment procedures in the last two decades; the bad news is that there is still a lot of work to be done. My lab's data indicate that the various procedures for assessing CU traits in youth do not agree very well with one another. Youth who score high on self-report measures do not necessarily score high on expert-rated measures or parent ratings. We also found that the more severe the child scored on the expert-rated measures, the less likely we were to receive back reports from parents. The parents of children high in CU traits often did not return the test to the researchers. Sometimes the apple does not fall far from the tree.[9]

Ideally, we would like the different assessment procedures to converge on very similar answers. This would help to develop a more coherent set of research findings and move the field forward more quickly. Nevertheless, the diversity and explosion of research in this area has led to a number of very important discoveries.

Among them, studies have shown that CU traits are fairly stable from childhood into adolescence and into early adulthood.[10] Indeed, CU traits appear to be more stable than conduct disorder traits (i.e., impulsivity, acting out) over time. More important, CU traits predict pretty well which children will go on to be convicted of crimes as adults.[11]

Research has even shown that CU traits assessed in boys between the ages of seven and twelve predict their psychopathy score when they reach age nineteen.[12] And Lyman and colleagues (2007) reported measures of CU traits at age thirteen that predicted adult measures of psychopathy at age twenty-four.[13]

The implications of this research have led psychologists to propose that *DSM-5* includes a child conduct disorder subtype that carefully assesses and includes CU traits. *DSM-5*, which was released in May 2013, has added a new conduct disorder subtype: *limited prosocial emotions.*

The new specifier of conduct disorder with *limited prosocial emotions* is the first attempt of the *DSM* to capture the relative importance of CU traits in the assessment of high-risk youth. Hopefully, this new subtype will capture these latter traits better than previous iterations of the *DSM*, but at the time of the writing of this book there was little research completed on this new *DSM-5* subtype.

In summary, research over the last twenty years has led to great advances in our understanding of how CU traits manifest themselves in children. We are moving rapidly towards being able to identify children who are at the highest risk for developing lifelong personality disorders. By getting the diagnosis correct, we can avoid incorrectly labelling such children as having ADHD or bipolar disorder or some other malady. As this research continues, we may be able to find ways to manage and perhaps even treat these children.

Neuropsychological Tests and Callous and Unemotional Traits

In addition to using scales or instruments to measure CU traits, psychologists and neuropsychologists have developed tasks or games to try to examine the systems of the brain associated with these symptoms. One game or task that researchers employ is called the *emotional lexical decision task*. A lexical decision is like a spelling test. Strings of letters are briefly flashed on a computer screen, and the participant has to decide if the letters form a real word or whether it is a nonsense or misspelled word. When the letter string forms an emotional word ('hate', 'kill', 'die'), people are faster to respond than if the word formed by the letter string is a neutral word ('table', 'chair', 'arm'). Processing emotional words taps into a brain system that makes us recognize the word very quickly. It turns out that the brain system believed to be giving the 'boost' is the amygdala – the amplifier in the brain that helps us pay attention to important salient events.[14]

The emotional lexical decision task was first used to study adult male psychopaths in prison. In a paper published in 1991, graduate students Sherrie Williamson and Tim Harpur, along with their mentor, Dr Robert Hare, found that, unlike normal people, criminal psychopaths failed to show faster responses to emotional words than to neutral words. This was seen as evidence that psychopaths 'know the words but not the music'. In other words, psychopaths know what 'love', 'hate', and 'kill' mean, but they don't *feel* the affective impact these words convey.

As an undergraduate student at the University of California, Davis, I had a mentor, Dr Debra Long, who was a professor who studied how the brain processes language. When I told Dr Long about the emotional lexical decision finding in adult psychopaths, she became interested in using this task to study whether university undergraduates would show deficits. So we used the Levenson Self-Report Psychopathy Scale (LSRP)[15] to examine whether self-report psychopathy scores would correlate with response times on the emotional lexical decision task in UC Davis students. We found that they did. The higher the university undergraduates scored on psychopathy, the worse they were at processing emotional words. We also found that high LSRP scores predicted which students would cheat on tests and encounter other student problems on campus.

Subsequent to these findings, researchers have shown that youth with CU traits show impairments in emotional lexical decision and related tasks.[16] In other words, CU traits in youth (and in adults) are linked to abnormalities in how the brain processes these affective stimuli.

A number of additional studies and tasks that have been developed over the last decade or so show that youth with CU traits and adults with psychopathy have deficits in processing emotional stimuli. These results are consistent with letters from the parents of such kids who ask if their children have an emotional learning disorder.

But as good as these tests are at teaching us what's wrong in these children, they are at best proxies for what is really happening in the brains of these children. To continue to examine in more detail how CU traits are manifest in the brain, researchers need to study the brain itself.

Brain Imaging in Children with Callous and Unemotional Traits

As I discussed in Chapter 4, the latest advances in MRI techniques permit scientists to study the brain in action. The functional MRI procedure is completely noninvasive and can be used to study children and adolescents. For example, we can put youth in an MRI scanner and show them pictures depicting emotional situations (i.e., results of a car crash) or pictures of neutral scenes (i.e., a car parked in front of a building). By mapping the brain's response to these pictures, we can show that it engages very differently for emotional pictures than it does for neutral pictures. In this way we can probe how the emotional systems of the brain are working. Indeed, scientists now can map out pretty much any process in the brain with this technology. Reading ability, mathematics, decision making, even moral decision making, are but a few of the kinds of processes in the brain that can be studied with fMRI. This technique has caused a revolution in the field of neuroscience in general, and the field of psychopathy in particular.

Children, however, present with some unique difficulties when you try to measure their brain function in an MRI. First, they need to lie still for about an hour, something that many children have difficulty doing. Children who are impulsive and agitated are even more likely to move around in the MRI scanner. And moving around in an MRI scanner while we are trying to collect data effectively blurs out the pictures of the brain and prevents usable data from being collected.

Childhood is also a time of rapid neuronal development, which can complicate scientists' ability to understand or interpret differences between two groups of children. Sometimes it might not be possible to know if an abnormality is just a delay in development or a true deficit that might last a lifetime.

At the time of the writing of this book, only about a half-dozen brain imaging studies have been published on children with elevated CU traits. But interestingly, these studies are finding pretty similar results. In one study, researchers found that the children with CU traits failed to activate the amygdala while viewing fearful faces.[17] The failure of the amygdala – an important part of the brain's emotional system – to engage when processing fearful faces may be part of the reason why children with CU traits are so aggressive. If two children are fighting, and one displays a petrified, fearful face, it is likely the fight will stop. But if the child with CU traits gets no signal or input from his or her amygdala when viewing the other child's fearful face, the fight may continue, and even escalate, with more severe injury as a result. The child with amygdala deficits might also get into more fights than other children.

Another area of the brain that appears to be abnormal in children with callous conduct disorder is the bit of the brain right above the eyes, the orbital frontal cortex. This very important part of the brain is responsible for determining a lot of our personality. Several studies have shown that callous conduct disordered children show impairments in the engagement of the orbital frontal cortex during learning tasks.[18] Among the implications from this research is that children with CU traits do not learn from punishment in the same way as other children. In other words, punishment, or even the fear of punishment, is not restraining the behaviour of these children in the same way it does for other children. The children with CU traits simply do not learn from punishment.

This work may have profound implications for how children with CU traits are managed. It also may have major implications for how parents raise these children.

Ivy League Lessons

The year was 2000, and my time in British Columbia was coming to an end. My main thesis adviser, the father of modern research in psychopathy, Dr Robert Hare, had retired from the University of British Columbia, and my postdoctoral supervisor, Dr Peter Liddle, had decided to move his family to Nottingham, England. I was on the job market. I'd sent out a dozen applications for academic positions. I prepped and practised my job talk for upcoming interviews. At the time I was unclear whether I was going to do another postdoc or whether I should try to go straight for a faculty position. I was just testing the water to see what was out there.

Over the next few months, I interviewed at the University of Wisconsin, the University of Florida, Yale University, and the University of Connecticut Health Science Center. All expressed interest in hiring me. At Yale the faculty had wanted me to take a position conducting brain imaging studies of posttraumatic stress disorder (PTSD). Yale was committed to researching and developing a better understanding of the brain processes involved in that condition.

Interestingly, the brain systems involved in PTSD are nearly the same as those involved in psychopathy, only in the opposite direction. Individuals with PTSD show exaggerated brain activity in the amygdala and anterior cingulate in response to salient stimuli. Psychopaths show the opposite — reduced activity. I thought it might be a good match at Yale. I could do a few years studying PTSD and then see if I could create opportunities to continue my psychopathy research. The forensic psychiatry programme at Yale was top-notch, and the professors expressed a great deal of interest in my brain imaging findings on psychopaths.

The University of Connecticut Health Science interview was for a postdoc with Dr Vince Clark. I'd met Vince years earlier when I showed him my weird psychopath brain waves at a scientific conference. Vince had himself studied brain waves for years. Vince had just received a new grant from the National Institute on Drug Abuse (NIDA) examining the P3 brain wave and fMRI activity in substance abusers. With his collaborators at the University of Connecticut, he found that the amplitude of the P3 brain wave predicted which individuals with substance abuse would relapse to drugs. It was a curious finding — the smaller the P3, the more likely the individual was to relapse to substance abuse. I wondered whether the P3 would predict which convicted felons would go on to commit other crimes.

Vince and I had a great interview. But as we were wrapping up, he told me that as much as he wanted me to come do a postdoctoral fellowship with him, I was doing extremely well so far in my career and I should take a permanent faculty position if I could get one.

Then he asked me to give another lecture while I was in Connecticut. There was a group forming a new research centre and while I was in town I might as well go talk with them. The lecture was to take place at an organization I'd never heard of before, the Institute of Living.

In my hotel room I went online and found out that the Institute of Living (IOL) was the third-oldest asylum in the United States. It was located on a large campus in the middle of Hartford, the state capital.

The next day Vince picked me up at my hotel and drove me over to the IOL for my noon talk. The grounds of the IOL were immaculate. The trees were blooming in all their autumn glory, and the brick hospital buildings provided the perfect backdrop. Vince told me that the grounds were designed by Frederick Law Olmsted, the landscape architect who designed Central Park in New York City. Rumour had it that Olmsted designed the grounds while he was a patient at the IOL.

The IOL has a rich and long history in psychiatry. The famous psychiatric patient HM (H.M. were the patient's initials) had been treated here. HM had suffered from intractable epilepsy, and surgeons had performed a radical bilateral resection of his temporal lobes to try to treat his seizures. The surgery removed the left and right sides of HM's brain that controlled memory; it helped to alleviate his seizures, but HM developed a unique memory problem. He could no longer encode new events. If you met him, talked to him, and then left and came back fifteen minutes later, he would not recall having met you. HM never recovered the ability to encode new memories. Apparently, he lived just around the corner from the IOL in a West Hartford retirement community.

When I entered the auditorium, I found four people eating their lunches – Vince and three very senior-looking men.

Vince introduced me to them: Drs John Goethe, Les Silverstein, and Harold Schwartz. All were in their mid to late fifties and seemed far more interested in their lunches than in my lecture.

It was far and away the smallest audience I had ever been asked to speak to. Nevertheless, I fired up my laptop and gave them my lecture on what I thought neuroscience was going to do for psychiatry in the next couple decades.

Vince had encouraged me to go big, so I laid it on a bit thick, talking about how neuroimaging might change how psychiatrists diagnose and treat patients with mental illness. And then I hit them with my brain imaging data on psychopaths.

As I finished, Dr Schwartz started clapping, and then the rest of them joined in. They huddled up, and then Dr Schwartz whispered something to Vince. Vince motioned for me to pack up my bag. We proceeded downstairs to the cafeteria, where I grabbed a bite to eat. Vince said I was to meet privately with Dr Schwartz following lunch.

I peppered Vince with questions about what was going on, but Vince said that he had been sworn to secrecy.

'Do they have an MRI?' I asked.

'No,' he replied.

'Do they have any neuroscientists?'

'No, nobody else here is trained in neuroscience.'

I had no idea why I was here.

Vince and I wandered the grounds for a bit. I had to admit the landscape was stunning.

When it was time for me to see Dr Schwartz, Vince took me over to the main administration building.

Dr Schwartz's executive assistant, Ruth Black, directed me to the waiting room outside Dr Schwartz's office.

A short time later, Ruth's phone buzzed. When she picked it up, she nodded and said, 'Dr Schwartz will see you now.' She escorted me through a set of double doors and into a large round office that looked as if it had been modelled after the Oval Office in the White House. Dr Schwartz sat behind a large desk. Behind him, the windows looked out onto the manicured lawns of the institute. There was a seating area with two sofas across from each other and two comfortable chairs at either end. Dr Schwartz stood. We shook hands and then he gestured to one of the sofas.

I thought briefly of lying down on the proverbial psychiatrist's couch, but decided Dr Schwartz might not share my sense of humour.

'You are probably wondering why you are here,' said Hank, as he asked me to call him.

I answered politely, 'Yes. I am interested to know what all the secrecy is about.'

'The IOL is going to make a big investment in research, and we have been interviewing individuals to find out what area of psychiatry to target. We have considered genetics, neuroimaging, and a few other areas. Your talk today convinced us that neuroimaging has the potential to make enormous advances in diagnosing and treating mental illness.' Hank paused for effect as I processed the details of 'making a big investment'.

He went on. 'Also, I am a forensic psychiatrist by training, and I really like your work in psychopathy. We might be able to help you do that work here since we hold the contract for all forensic mental health in the state in collaboration with the University of Connecticut Health Science Center.'

Interesting, I thought.

'I've decided to partner with the Department of Psychiatry at Yale University in the recruitment of a director for the new research centre we are creating. We have forged a relationship between the IOL and Yale and are planning to offer Yale faculty lines, funded by the IOL, to create a world-class psychiatric centre.

'We have been interviewing potential directors for a while, but I have not found one who has inspired me or my committee. We have interviewed only a few select individuals and we have not advertised the centre or our plans.'

So that's why all the secrecy. They were stealth-interviewing candidates to direct the centre.

'How is this funded?' I asked.

'A number of families have donated some money for this endeavour. One bequest was quite detailed and quite large. The donor wanted a research centre to be formed to help improve mental health care in general and here at the IOL in particular. The IOL has a long and rich history of caring for the mentally ill. We have been focused on clinical training for much of the last hundred years and are planning to start doing more research again.' Hank was referring to the three-hundred-year history of the institute.

'Do you know the name Dr Henry P. Stearns?' Hank asked.

'Yes,' I replied. 'He was a psychiatrist who testified about moral insanity at the trial of Charles Guiteau in 1881.'

'I'm impressed,' said Hank. 'I actually didn't know that Stearns testified in that trial, but he was a famous forensic psychiatrist who ran this institute when it was called the Hartford Retreat for the Insane.' We went on to discuss Stearns's legacy at the IOL.

Then Hank turned to me and said, 'Dr Kiehl, I'd like to offer you a position here at IOL. I want you to help me build a neuroscience centre, recruit the right director and other junior faculty. I've already talked to the chair of the Department of Psychiatry at Yale and they will offer you a faculty line. Apparently, they want you for another position and figured you would help them out in that endeavour as well.

'So I need to know your terms,' Hank continued. 'What do you want in terms of salary and start-up package?' And with that he stood up and headed to the door.

'I'll give you a few minutes to think about it,' he said, and left me alone in the room.

My mind nearly exploded. On the one hand, it sounded like a great opportunity. On the other hand, I would have to build a programme from scratch. That might take a long time. At this point in my career, I needed to spend my time writing and publishing. My mind was racing.

I thought about my salary. I wasn't sure where to start. I knew what salary my friends were being offered by other good schools. And I knew that start-up packages for new faculty ranged dramatically in size and scope. Start-up packages could include salaries for research staff, money for computers and equipment, and other expendables.

I decided to just throw out some big numbers for my salary and start-up package at Hank and try to buy myself some time while he checked with his committee.

I'd never been in a situation like this. During my past interviews, I was just dropped off in front of Human Resources, where they talked to me about salary ranges and benefits. There wasn't really that much to negotiate.

After a few short minutes, Dr Schwartz returned and sat down. He gently laid a folder on the table in front of me and then asked me if I had any questions.

Yes, I thought to myself, *I have hundreds of questions*. But I instinctively trusted this man. I wasn't sure why. It was the complete opposite of the feeling I had when interviewing psychopaths.

'Are you planning to buy an MRI scanner?' I asked.

'Yes. I am going to ask you to go buy one.'

My brain fizzled. Nobody asks a junior faculty member to buy an MRI.

'I need someone who understands the engineering of the system.

I need someone who can figure out what system will be the best for the next decade or so,' Hank said.

'Where are we going to put it?' I asked.

Hank got up and retrieved a large set of architectural drawings. He laid the plans of a large two-storey building on the coffee table in front of us.

'I want you to redesign this building for the new research centre. And you get to pick where the MRI will go. This building has been vacant for over seventy-five years and it has fallen into disrepair. It's an 1870s English Tudor revival–style building that we want to completely gut and renovate. I want you to design the research centre.'

I looked at the plans. I didn't even know what *English Tudor–style* meant.

Hank patiently waited for me to process the information he had provided.

He continued, 'You designed the new MRI centre at the University of British Columbia and helped them get a grant to fund their research scanner.'

How did he know that? I wondered. Then I realized that my old supervisor, Dr Liddle, whom I had helped write the grant and do the plans for the building, had written a letter of recommendation for me. Hank must have got hold of the letter from Dr Clark.

Hank went on. 'You are the right person for this job. And I like your enthusiasm and work ethic.' Hank confessed that he had called my former supervisors at UBC and talked with them while I was at lunch. Apparently, they had told him about my long days, nights, and weekends in prison. They had also emphasized that I was a tech guy who liked to play with MRIs.

'So what's it going to take to get you to sign the papers?' He motioned to the folder he had set on the desk in between us.

'I want you to leave this office knowing that you got a good deal, and I want to get on with making this research centre a reality.'

'Well, if you are paying for the MRI and renovating the building, I guess I don't need to have any of that included in my start-up package. All I need are funds to outfit and staff the MRI and start my research programme.' I added, 'This MRI-compatible equipment is not cheap. It will take several hundred thousand dollars to set up all the hardware and peripherals for the MRI suite. And then I would need to have a couple of staff to start my projects.'

Hank nodded slowly and then asked, 'What do you want for a starting salary?'

I made up a number, twice what all my friends were being offered for starting faculty positions at good universities.

'Okay,' Hank said. 'This is what I am going to do. I'm going to give you a little more than you asked for your salary. And then I'm going to triple your start-up package.'

I was unable to breathe. I started looking for a pen so I could sign those papers before he changed his mind.

'I really need you to come here and help me get this programme up and running. When can you start?'

'Tomorrow,' I said honestly.

Hank had a long laugh and then smiled at me. 'Why don't you finish up what you are doing in Vancouver, and then we will have you come out a couple times to meet with the architects. You can start to negotiate the MRI, and I'll pay you as a consultant for that work. Then you start here full-time in July.' The start date was more in line with the academic calendar.

'Okay,' I managed to mumble.

And I signed the documents.

Ruth magically appeared and whisked away the paperwork.

'Now let's celebrate with a drink and you can tell me more about your work with psychopaths. I'm fascinated.'

We spent the next hour discussing my research and Hank's vision for the research centre.

Vince Clark was smiling when I emerged from Dr Swartz's office. 'How did it go?' he asked as we descended the stairs. 'Did he make you a *say no* offer?'

'What's a *say no* offer?' I had never heard the expression before.

'It's an offer you can't refuse, one that takes all the others out of contention. That's what Hank said he was going to do.'

Ambushed. Well, it was the best ambush anyone had ever pulled on me. I couldn't be too angry with Vince.

'Yes, he made me an offer I could not refuse. I am moving to Connecticut,' I replied and then it kind of struck me.

I was leaving Vancouver, my home for the last seven years. I recalled the first day I drove across the border in my little Toyota pickup truck loaded down with everything I owned, filled with excitement about starting graduate school. How quickly the time had passed.

That feeling of unbounded energy and excitement I felt then was back. Gone was the uncertainty about where I was going to go and whether or not I would be able to get a job. I was on to the next adventure in my life.

I returned to Vancouver and notified my landlord that I would not be renewing my lease next summer. I decided to take some time off between moves and hike part of the Pacific Crest Trail in the late spring.

I also started researching MRIs to buy and sought out advice from Richie Davidson, whom I had interviewed with at the University of Wisconsin, and who had wanted to hire me to work at the new brain imaging centre he was getting funding for at the Madison campus. He understood my decision to take the job in Connecticut, telling me the IOL/Yale job was a great opportunity, one I could not pass up. Richie gave me lots of tips about buying an MRI. He told me to ask all the companies that sell MRIs to send their sales people over to meet with me. He suggested I have them all come at the same time and make them wait outside my office, signalling that they would all be competing against one another.

So I arranged to have representatives from GE, Phillips, and Siemens visit at the same time.

I had the Phillips rep come in first. He showed me the latest plans for their first attempt to break into the research MRI market. When I asked him how his system was different from the other vendors', he didn't have a good answer. He just claimed that he could beat any price because they were really trying to break into the market in Connecticut. Price was a secondary concern; our main objective was to get the best hardware. Phillips was off the table.

Up next were site visits for Siemens and GE. The best site for Siemens was Massachusetts General Hospital in Boston; and GE offered to fly Hank and me to their headquarters in Wisconsin to show off their production facility.

Siemens had a head-dedicated MRI, the Allegra, that was state of the art, German built, and the current rock star in the brain imaging field. I was looking forward to testing it out. I also learned that the IOL had just partnered with Hartford Hospital, which had GE MRI systems in their radiology department. Siemens was interested in breaking into the market in Hartford; the price of the Allegra started to drop to make it more competitive. In the end, the price dropped by more than 30 percent, over a million dollars off the list price. Siemens really wanted the sale.

Siemens arranged for us to visit their flagship research site in Boston. I e-mailed my pal Larry Wald, a physicist, and gave him a heads-up that we were visiting. Larry is easy to spot in a crowd. He is a rather large man, standing well over six feet tall, with bright ginger hair. Larry shares my penchant for steak, and he is married to a vegetarian, so he is always willing to get out for a ribeye. Larry was the best physicist I had ever met — brilliant technically and a pleasure to work with.

Larry gave us all the nitty-gritty details of the Allegra system. He wowed us with the technical advantage, but he also told us about some of the problems. I could feel the Siemens sales rep cringe when Larry told us about the fact the system runs so fast it is hard to cool it down. I took notes in clear view of the Siemens sales rep. I planned to use these details to negotiate a fix on our system — and perhaps even lower the purchase price. Nevertheless, Siemens was on the top of the short list.

It was a great visit capped off with a lovely steak for Larry at Morton's Steakhouse.

Now it was time for one more visit.

Hank and I went to visit GE. The company flew us to Wisconsin and showed us their plans for the next generation of research MRIs. We were wined and dined.

My friend Bryan Mock met us and he helped set up GE's presentation in their lecture hall. A couple of other guys were running the show, and they told us all about the latest GE hardware and software.

I asked them about the software bug that allowed us to collect only 512 images in succession before stopping and restarting. They hadn't fixed that problem yet. The salesmen then showed us a bunch of new options and data processing pipelines. I asked questions and pointed out limitations and potential flaws in their system and pipelines. The salesmen were clearly not technically oriented. I noticed Bryan crack a smile from his seat in the corner. Their design was geared towards a turnkey clinical system, and it was not research friendly.

I was probably a little hard on the salesmen, but I was trying to prove to Hank that I was the right person for the job and that I had done my homework.

As we left GE's headquarters, I walked out with the vice president for MRI development. I told him that Bryan Mock knew how to fix all the problems that I raised during our visit. I suggested he promote Bryan to run the development unit – that Bryan would make GE competitive again in the functional brain imaging field But for me, GE was off the table. Their technology at the time was just not competitive.

Hank and I regrouped in his office the beginning of the following week. I presented my argument for purchasing a Siemens Allegra. I showed Hank the original list price and then the final negotiated price. I'd also had Siemens guarantee the delivery date of the system to coincide with construction of our new research building. I'd negotiated all the finer points – stability testing, up-time guarantee (with penalties for noncompliance), and service contract terms and price. All in all, I had tapped the expertise of nearly every director of an MRI programme in the country. I had wanted to make sure we got a good system and a good deal.

Hank was pleased, and we ordered the Allegra. Hank mentioned that I had saved over one million dollars off the list price. A week later a very nice bonus showed up in my pay cheque. I was going to like living in Connecticut.

Birth of the Olin Neuropsychiatry Research Center

One of the main donors to the Institute of Living had been the Olin family, a family who had had a number of children treated at the IOL. Dr Hank Schwartz had decided to name our new research building in their honour. Remodelling the Tudor revival–style building was my next job.

The preliminary plans had the building organized like a doctor's office, with patient waiting rooms outside each office. I changed the design plan to include only one central waiting room near the ground-floor front entrance. The reorganization added about five additional offices, for a total of fourteen offices on the first floor. Then I noticed there were no toilets on the first floor, so we fixed that. Patient rooms were added so we could conduct interviews and testing.

I worked with an acoustical engineer to make sure the MRI suite was soundproof. The design proved to be so quiet that visitors to the control room would often ask if the machine was working. The MRI techs were happy they didn't have to listen to the constant beeping the MRI makes while it is collecting data.

During the process of buying the MRI and redesigning the research building, I also helped Dr Schwartz recruit a centre director. Dr Schwartz wanted to hire a senior psychiatrist who had a breadth of experience but a focus on schizophrenia. After reviewing a number of candidates, we selected Dr Godfrey Pearlson from Johns Hopkins University. Godfrey was known for his studies of brain structure in schizophrenia, and he wanted to expand into functional MRI and other brain connectivity measures.

Godfrey also asked us to interview Dr Vince Calhoun, who would join the centre as an assistant professor. Vince was the electrical engineer I had met via e-mail many years earlier when we were both trying to solve data collection problems we were having with our GE scanners. He was developing new ways to analyse brain networks using functional MRI data. Vince would be a fantastic hire.

I recommended to Hank that we hire them both, and we did. The core of the centre was formed.

Godfrey was an experienced researcher, and he quickly filled the Olin Center with new staff and postdocs. We started writing grants emphasizing the unique and untapped clinical resources of the IOL. The inpatient wards at the IOL had more than three thousand psychiatric admissions per year. It was a clinical gold mine.

I formed my research team and started writing grants to be submitted to the National Institutes of Health to fund my research aims. For my psychopathy work, I planned to recruit psychopaths while they were on probation or parole in the community. In schizophrenia I was interested in one main goal. I wanted to know if brain imaging could be used to help diagnose the condition at early phases, perhaps even before the onset of serious symptoms. This advance might aid in the diagnosis of the condition, but critically, early treatment might minimize the tragic progression of the disease.

Dr Peter Liddle, one of my graduate mentors, had hypothesized that functional brain activity could be used to help predict whether youth would develop the illness. I brought the ideas Dr Liddle had started to the centre and sought to develop new paradigms to address the problem of early diagnosis in schizophrenia. In addition to the new paradigms, I was also armed with Dr Calhoun's new algorithms for analysing brain imaging data. Together we had created a powerful team of researchers, and we soon started to win grants from NIH to pursue our ideas.

Concrete Psychopaths

Alongside my grant writing, I continued to analyse the prison data from graduate school.

A couple of the studies I conducted before leaving Canada had their roots from insights I gleaned through my clinical experiences with the inmates. After I had finished my master's thesis, I was asked by the director of the psychology clinic at the Regional Health Centre Prison, Dr Carson Smiley, to sit in on group treatment classes for the violent offenders and sex offenders. The Canadian Department of Corrections allows master's-level psychologists to serve as clinicians. Dr Smiley wanted me to learn more about the group therapy they were conducting so that I might be able to contribute to improving it. He also wanted to know if the neuroscience I was conducting might give them some insights into the deficits psychopaths experience. The goal was to use my neuroscience findings to help develop better treatments for psychopaths.

At the time, I really didn't know much about treatment for violent offenders or sex offenders, but I figured it would be a good experience, so I agreed to help. The psychiatric nurse ran the therapy sessions, and I sat in the room and observed.

The therapy sessions employed cognitive behavioural therapy, or CBT. CBT is a type of talk therapy that tries to get people to look at problems from a different perspective, using a goal-oriented, systematic approach. With violent offenders and sex offenders, CBT is used to challenge individuals on their maladaptive thinking patterns and beliefs. The goal is to replace these maladaptive errors in thinking with more constructive patterns that will reduce future antisocial behaviour.

I noticed in these group therapy sessions that psychopaths could not grasp abstract concepts. For example, the nurse would often use metaphors to emphasize her points. In more than a few cases, the psychopath she was speaking to just stared back at her with a blank expression.

I had seen that expression on the faces of psychopaths before. It was a *tell* – just like in poker. The facial *tell* was apparent when I would ask psychopaths about whether they worried about things. The classic example of the type of worry I am referring to is the kind that occurs in individuals who suffer from obsessive-compulsive disorder (OCD). Individuals with OCD worry all the time about things; for example, being concerned that they have left the oven on every time they leave the house. Eventually, it can escalate to the point where people with OCD won't leave the house because they are so worried they left the oven on and they have to constantly go back and forth to make sure that it is off.

If you ask a psychopath if he or she has ever worried about leaving the oven on after leaving the house, you will get this blank expression I am referring to. Psychopaths have no concept of what worrying like this is like. Psychopathy and OCD are at opposite ends of the spectrum. I've never met a psychopath with OCD — I don't believe one has ever existed.

I was fascinated to see that psychopaths were unable to appreciate metaphors. It seemed that psychopaths were able to understand concepts only if they were presented in concrete terms. I quickly designed a series of studies to examine whether the psychopaths' inability to process abstract concepts generalized to more simple stimuli. My first study examined how psychopaths process abstract and concrete words.

Psychologists break down words in a number of different ways. They can be broken down by their frequency of usage in the population. For example, a low-frequency word would be *aardvark* and a high-frequency word would be *deer*. It turns out your brain processes common words differently than uncommon words.

The brain also differentiates abstract words from concrete words. Abstract words are ones that don't have any physical meaning. For example, *finite* is an abstract word, while *table* is a concrete word. I developed a database of concrete and abstract words that were matched on things like frequency of usage, word length, and number of vowels. I wanted to know whether psychopaths differentiated concrete from abstract words under different conditions. I made sure that all the concrete and abstract words were neutral in affect. I didn't want my studies of abstractness to be confounded by the emotional processing deficits in psychopaths.

But I also wondered if the psychopaths' emotional abnormalities might be due to deficits in processing abstract information. After all, emotional words are often abstract. 'Love' for example, is a highly abstract word. It means many different things to people, and it doesn't necessarily have a concrete meaning. People use poems, musicals, and theatre to describe the abstract aspects of love — but what does love really *feel* like?

I always ask psychopaths during interviews, 'What does love mean to you?' The most common answer is 'sex'. That answer is often followed by the psychopath describing his favourite sexual escapades. Psychopaths get stuck on the physical, the concrete, and they fail to describe the abstract connections that love provides.

In my first studies, I used brain waves to examine the neural response in psychopaths to abstract words. I found that psychopaths did not show the same brain wave differences between concrete and abstract words as nonpsychopaths and healthy individuals did.[2]

The brain is truly amazing at differentiating different word types. One hundred and seventy-five milliseconds after the word is presented, the brain has started to put abstract words on a different processing path than concrete words.[3] But for psychopaths, brain waves show that all words were processed in the same way, going down the same path. Their brains did not respond differently to abstract and concrete words.

It seemed pretty clear from these brain wave studies that psychopaths were impaired in processing abstract words; something was wrong in the neurocircuitry of their brains that prevented the two types of words from being processed differently.

Next I used the fMRI technique to find out *where* in the brain things were going wrong in psychopaths while they processed abstract words. In nonpsychopaths, the right anterior temporal pole is the critical node for differentiating abstract representations of speech, including the difference between abstract and concrete words (see Figure 5). As predicted by my research, psychopaths showed huge deficits in this region of the brain when processing abstract words.[4]

The temporal pole is the area that integrates information from our senses: like integrating video and sound together to make a film. One way to think about the deficits in psychopaths is that they have problems putting the abstract content in the story line.

The research suggested that the reason psychopaths failed so badly in treatments that emphasized abstract concepts was that their brains were unable to process them. If psychopaths were going to benefit from treatment, the neuroscience data were telling us that the treatment must be presented in simple, concrete terms.

My PhD research studies from Canada had indicated a number of brain regions were impaired in psychopaths: the amygdala, hippocampus, anterior and posterior cingulate cortex, and the temporal pole. Other scientists had shown that the orbital frontal cortex was also impaired.

A summary of the regions believed by the start of the twenty-first century to be implicated in psychopathy is outlined in Figure 6.

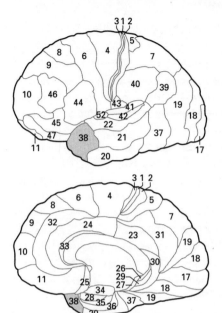

Figure 5. Lateral (top) and medial (bottom) illustrations of the human brain. The medial view is as if you sliced the brain down the middle and pulled the two halves apart to look inside. The numbers represent areas defined by a labelling system developed by Dr Brodmann in 1909. Scientists use the Brodmann numbering system to help them compare results across studies and across laboratories. The results from the abstract word processing studies implicate the anterior temporal pole in psychopathy (Brodmann area 38).

Figure 6. Summary of the brain regions believed to be implicated in psychopaths by the year 2000. Regions include the amygdala, hippocampal complex, anterior and posterior cingulate, anterior temporal pole, and the orbital frontal cortex. The numbers represent different areas of the brain as defined by the work of Dr Brodmann. See prior figures for complete details on the illustrations.

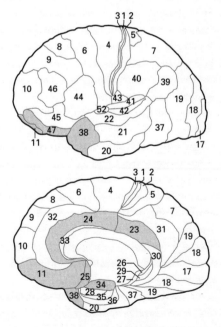

Little Green Maps

One of the great benefits of being at Yale was that there were top experts in the world in nearly every topic imaginable. I immersed myself in the Yale community and continued my education as a scientist, attending all the speaker lectures in psychology, psychiatry, neurology, neuroscience, and economics.

I also started a new quest: to work out how all the pieces of the psychopath's brain puzzle fit together. I began reviewing different theories of brain development. The burning question was, *Why do psychopaths have abnormalities in so many different regions of the brain?* Was there any logic to it? Was there any way to create a unified theory around the results that I, and others, had published to date?

The answer to this puzzle started to form when Dr Hilary Blumberg of the Yale Psychiatry Department gave a seminar on brain correlates of depression. Hilary started her lecture using a map of Brodmann areas. Korbinian Brodmann (1869–1918) was the physician who painstakingly developed an anatomical classification system for different brain regions, classifying brain systems by the type and density of different neurons that occupied these regions.

I was familiar with Brodmann areas as an anatomical labelling system, but Hilary put up a figure that I had never seen before. It showed the Brodmann maps of various brain systems in different colours.

In blue were the basic sensory systems – vision, hearing, and motor. These areas had similar types and densities of neurons that generally performed the same functions – processing basic sensory inputs. The colour yellow indicated higher-order sensory cortices. Pink signified the prefrontal and parietal cortex. The larger prefrontal cortex in humans is what differentiates us from other primates. And then finally in green were regions of the brain Brodmann had classified as paralimbic. The paralimbic regions included the classic limbic structures: the amygdala, hippocampus, and anterior and posterior cingulate. But it also included the orbital frontal cortex, insula, and temporal pole (see Figure 7).

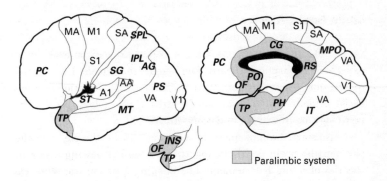

Figure 7. Brodmann's maps of the different systems in the brain based upon how neurons are organized. Notice the similarities between Brodmann's paralimbic circuitry (the shaded areas) and the areas neuroscience had shown are abnormal in psychopaths in Figure 6. The two figures are dramatically similar, a fact that led to the development of my paralimbic dysfunction model of psychopathy.

It was at that point that a lightbulb went off in my head. *That's it,* I thought! The brain regions I had found to be abnormal in psychopaths mapped directly onto the regions Brodmann labelled paralimbic. It was uncanny how much the two maps overlapped. I begged Hilary for a copy of her slide, and then I went straight to the library to read everything I could about this paralimbic circuitry Brodmann had described.

A year of research – and reading thousands of articles – followed. It was a monumental effort, as I laboured over weekends and evenings to pull together a comprehensive picture of psychopaths' brains. As a result, I developed what I called the *paralimbic dysfunction model of psychopathy,* which I published in a peer-reviewed neuroscience journal.[5] My original draft of the paralimbic dysfunction model of psychopathy was longer than my PhD thesis.

But the complete picture came together like a meal with the perfect bottle of wine. My model included converging evidence from the emerging field of brain imaging in psychopaths, reanalyses and reinterpretation of brain wave abnormalities in psychopaths, and finally, a detailed review of what happens to people following brain damage to certain paralimbic regions. It turns out that if a person has a stroke or some other injury to a part of the paralimbic system of the brain, it may precipitate psychopathic symptoms.

The Famous Patient from Vermont

Most people know that if you get knocked unconscious, you can develop some problems remembering things. Amnesia, or memory loss, is a common symptom following concussions. However, damage to the brain can lead to a vast array of other types of problems. Sometimes brain damage can even lead to changes in personality that mimic what we see in psychopaths.

The most relevant neurological case study in this regard is that of Vermont railway worker Phineas Gage.[6] On 13 September 1848, Gage suffered a penetrating head trauma when a 1.2-inch-diameter iron rod was accidentally blasted up behind his left eye and out the top of his head.

Gage's case is notable for several reasons. First, he miraculously survived. Indeed, it has been reported that he never even lost consciousness following the accident. Second, Gage had a doctor who wrote about his case so that we could all learn from it. And finally, Gage was transformed by this accident from a responsible railway manager and husband to an impulsive, irresponsible, promiscuous, apathetic individual. Many of Gage's symptoms are consistent with those classically associated with psychopathy. The rod that passed through Gage's brain destroyed several parts of the paralimbic system.

Subsequent studies of patients with brain damage like Gage's suggest that the paralimbic system mediates most of the behaviours related to psychopathy. Neurologists originally named the condition Gage suffered from *pseudopsychopathy*,[7] but it was subsequently called *acquired sociopathic personality*.[8] In other words, if you damage a part of the paralimbic system, you can *acquire* a psychopathic personality.

As a group, paralimbic brain damaged patients are characterized by problems with aggression, motivation, empathy, planning and organization, impulsivity, irresponsibility, poor insight, and lack of behavioural controls.[9] In some cases, paralimbic brain damaged patients may become prone to grandiosity and confabulation.[10] These are all symptoms that we see in psychopaths.

Damage to some areas of the paralimbic system are not that uncommon. For example, when the brain is slammed forward against the front part of the skull, it can rub against the bony ridge that exists right above the eyes. This rubbing can damage the *orbital frontal cortex* of the brain. This is the type of injury that can occur in American football players who suffer repeated concussions. Whether due to a single event or to the cumulative impact of multiple head traumas, individuals who damage their orbital frontal cortex can end up developing problems just like Gage. It is this reality that is just beginning to be recognized by former National Football League players as a potential occupational risk to their sport.

As I put these pieces together, I realized that it might have been a blessing that my athletic career had been cut short at university by my knee injury. I'd already had four concussions in high school. Fortunately, my brain seems to have survived intact. I have to admit, though, every time I hop out of the MRI scanner after being in a research study, a part of me worries that my orbital frontal cortex might show up as damaged.

Gage did not have all the symptoms of psychopathy. My literature search clearly indicated that damage to just one or two parts of the paralimbic system does not lead to the full constellation of symptoms that we see in psychopaths. But damage to any part of the paralimbic system will often elicit one or more symptoms of psychopathy.

But even if multiple parts of the paralimbic system are damaged late in life, some symptoms of psychopathy don't appear. *Callousness*, for example, is one of the traits that is found in psychopaths but rarely found in adult patients with paralimbic brain damage. The reason? It may take years of paralimbic abnormalities in an individual for a callous streak to develop; most paralimbic brain damaged patients don't live long enough to develop this trait.

This reasoning is supported by a study of two individuals who, as infants, suffered brain trauma to the paralimbic system. The researchers indicated that as adults these two patients had a higher incidence of callous behaviour than is typically observed in patients who suffer similar lesions when they are adults.[11]

In other words, if psychopaths are set up from birth with paralimbic brain abnormalities, it might precipitate the development of callousness in them as adults.

Interestingly, paralimbic brain damaged patients sometimes also develop incontinence (i.e., bed-wetting). Earlier I pointed out that youth who experience chronic bed-wetting may be at higher risk for future violent behaviour. It remains unanswered whether the paralimbic brain damaged patients who develop incontinence are more violent than other paralimbic brain damaged patients. This is clearly an avenue that needs to be examined.

In addition to the similarities between the symptoms of psychopathy and the behavioural changes exhibited by patients following brain damage to the paralimbic system, there are also similarities in neuropsychological profiles between these two groups.

Neuropsychology uses tasks, or games, to understand the problems a person might be experiencing following brain damage. During my literature review, I discovered that the kinds of games patients fail following paralimbic brain damage were strikingly similar to the games psychopathic inmates fail.

Paralimbic brain damaged patients, like criminal psychopaths, have problems recognizing the nuances of voice or facial expressions. They also have problems regulating their behaviour, have poor decision-making skills, have trouble inhibiting their behaviour, appear apathetic, have trouble avoiding harmful situations, are promiscuous, show rebelliousness, tend to disregard social convention, and lack respect for authorities.

Damage to other regions of the frontal cortex (i.e., superior frontal or dorsolateral prefrontal), to the parietal cortex, or to the occipital cortex does not lead to these kinds of behavioural symptoms or cognitive abnormalities; that is, damage outside of the areas that Brodmann labelled paralimbic does not lead to psychopathic symptoms.

In other words, it appears that it was the paralimbic system that was failing in psychopaths.

And so all the pieces came together. I had found an answer in my quest to unravel the mysteries of the psychopathic brain. Armed with my new theory of psychopathic brain abnormalities, I started to design new studies to test specific components of the paralimbic system in psychopaths.

Ivy League Comfort

My life at Yale University and the Institute of Living was richly rewarding, both personally and professionally. I had developed a close group of friends with whom I skied, mountain biked, and went on holiday, adding some balance to my life after my workaholic graduate school life.

I was winning grants from the National Institutes of Mental Health and the National Institute on Drug Abuse to study psychopaths and substance abuse. And I'd been asked to serve on the NIH study section — a special privilege, where you are asked to review grants submitted by other scientists. I was climbing up the academic ranks.

But life was not without its hiccups. One frustration was that finding and then recruiting psychopaths for research was proving to be much harder in the community than it was in prison. Psychopaths, as one might imagine, often disappear once they are released from prison. And even if you can find them, they hardly ever show up for research appointments on time, if at all.

My laboratory tried everything to fix these problems. We sent participants postcard reminders, called them the week before, a couple days before, the night before, and even the morning of their appointment. But we still had lots of no-shows. My staff were spending most of their time trying to schedule appointments rather than actually collecting data.

On the rare occasions when individuals did show up for appointments, their appearance was often associated with some interesting histrionics. They would show up too hungover to be interviewed, and we would have to send them home. Or they would show up drunk or high on drugs. Several times we had to call security to escort a drunk and belligerent psychopath out the door and into a cab for a ride home. I got to know the security personnel at the hospital very well.

It was becoming clear that it was safer to work with psychopaths in prison than to work with them in the community.

Nevertheless, we persevered and continued to make slow but steady progress in scanning psychopaths in the community.

Female Psychopaths?

I gave a grand rounds presentation to my colleagues at Yale psychiatry on the latest research developments in psychopathy. Following my lecture, I was approached by a clinician who worked in a local maximum-security prison. The clinician asked if I would come evaluate a female prisoner for treatment amenability. I agreed.

The offender, 'Judy', was twenty-five years old. She had blond shoulder-length hair and a slim build, weighed just 105 pounds, and stood five feet three inches tall. I prepped for my interview with Judy by reviewing her institutional files. The files were several inches thick, and they also contained a number of videotapes.

As I got situated in the office the prison had prepared for me, I inserted one of the videocassettes into the VHS machine. The tape contained security footage of an interview room not that dissimilar from the one in which I was sitting. The video showed Judy on the opposite side of a large table from a psychologist conducting the interview. I watched the video for about ten minutes and then it happened – Judy launched herself across the table in a full attack on the therapist. Throwing haymakers, left and right, Judy landed two blows before the therapist knew what was happening. The therapist collapsed onto the floor, where Judy straddled her and got six or seven more blows in before security staff entered the room and pulled her off the therapist. Dark blood flowed onto the floor from the unconscious therapist's mouth – it appeared that the therapist had lost several teeth in the attack.

I rewound the tape and watched it again. Judy showed no warning signs; it was an unprovoked, vicious attack. Security must have been nearby since the therapist had had no time to hit the emergency call button; she had been completely disabled by the first two blows.

I shut off the tape and started reading Judy's files. Judy had been in and out of prison since the age of twelve. As a sixteen-year-old, Judy had shot and killed another youth. Judy's defence was that she did not know the gun was loaded when she pointed it at the other teenager's head and pulled the trigger. She received a sentence for involuntary manslaughter.

Judy had a long history of getting in trouble, and inside prison she was a nightmare to manage. She had dozens of fights with other inmates and had two other violent assaults against staff. The clinicians at the prison asked me to do an evaluation and determine if she met criteria for psychopathy. Following my evaluation, the treatment team wanted to know if I might have suggestions for any type of treatment that might help them control Judy's behaviour.

I completed my review of the collateral sources of information in Judy's files. I had evidence of serious emotional and behavioural problems in nearly all domains of her life. Even though I had interviewed hundreds of inmates by this point in my career, this would be the first female inmate I would interview. And I was hoping that it would not be the first interview that ended in a physical altercation.

Judy arrived at my 9 a.m. interview still half asleep. She groggily shook my hand and plunked herself down in the chair opposite from me. I explained to Judy that the interview was to help identify treatment options for her.

Judy's background was identical to that of the hundreds of male psychopaths I had interviewed. As a young child she failed to learn from experiences or punishment. Judy's parents responded to her aggressive behaviour and mischief with punitive sanctions, culminating with 'grounding' her for months at a time to the confines of their home. While stuck at home, Judy would sabotage electronics, using screwdrivers to mess with televisions, stereo equipment, and even the fire alarm. Indeed, setting off the fire alarm in the middle of the night became a bit of a pastime for Judy. She enjoyed watching her parents search frantically for any sign of smoke or flames.

Judy started carrying a gun at age fourteen, taking it with her just about everyplace she went. It was a Beretta semiautomatic pistol with a nine-round clip. I encouraged Judy to tell me about her firearm escapades. She regaled me with story upon story of her marksmanship and superiority with firearms relative to her peers. She also told me that more than a few times she had emptied the entire clip in drive-by shootings at rival gang members' houses. She would do her shooting sprees with one teenage boy in the car, a measure to ensure that if they were pulled over by the police, they could act as if they were just out on a date. The ingenious ploy worked more than once when they were stopped and questioned.

I turned the conversation back onto Judy's relationship with her parents and siblings. Judy had not seen her parents in a couple years, and she assumed her two younger siblings were doing just fine. Her siblings had not followed Judy down the path of antisocial behaviour.

As I was wrapping up the interview, I returned to talk about Judy's most significant index offence as a teenager – the involuntary manslaughter charge for killing another teen.

I asked Judy to explain what had happened.

Judy started to tell her rehearsed story about how she and two other teens were playing with a gun when she looked up at me and paused. I saw in her eyes a mix of anger and confusion. She realized that I had trapped her. I braced for her to attack me, but she just slouched down a bit more into her chair.

I had used our initial rapport to get Judy to talk about being an expert with that Beretta handgun prior to the 'accident'. She knew now that claiming the shooting was due to her inexperience was not going to fly given that we had established she was an expert with the handgun several years before the accident.

Sitting in the chair she said: 'Well, cat's out of the bag, isn't it?'

'Yes,' I answered.

'Then I might as well tell you the true story. It's not going to matter; I can't be charged again in that case.' Judy was referring to the fact she had pleaded guilty to involuntary manslaughter and she could not be tried again for the same crime.

'Michael dared me to pull the trigger, so I raised the gun to his head and pulled the trigger on an empty chamber. Click. I was just going to scare him. But then Michael dissed me again and said that he knew I didn't have the balls to shoot him. So I racked a round into the chamber and put it up to his head and fired. He's not dissing me anymore.' She looked at me with flat, emotionless eyes. They were the same eyes I have seen in hundreds of male psychopaths, but it was the first time I had ever seen them in a woman.

Judy stood up and said she was done with this interview. She left the room and went back to her cell.

My report to the prison would indicate that Judy was going to be very difficult to treat.

There is not a lot of research on female psychopaths. The best available evidence suggests that for every ten male psychopaths, there is one female psychopath. Like their male counterparts, female psychopaths tend to get in trouble with the law and often end up in prison. The rate of psychopathy in female offenders is 15 to 20 per cent, the same as the rate in male offenders. But only about 10 per cent of the prison population is female. Thus, female psychopaths are much rarer than male psychopaths. And there have been no published brain imaging studies of female psychopaths. Since my interview with Judy, my laboratory team has collected brain imaging data from female offenders, and we plan to publish the first neuroscience studies of female psychopaths in the near future.

Changing Environments

One morning I got a phone call from my old friend Dr Vince Clark. Dr Clark had left the University of Connecticut Health Sciences Center a few years earlier for a faculty position at the University of New Mexico. When I would see him at scientific conferences, he would tout the benefits of living in New Mexico – 320 sunny days a year, no humidity, affordable housing, great skiing in Taos Ski Valley, and world-class mountain biking outside your door in the local Albuquerque mountains. Dr Clark seemed very happy with the move.

He was calling to invite Dr Calhoun and me out to New Mexico to give talks at the Mind Institute on the campus of the University of New Mexico.* The Mind was a nonprofit world-class neuroimaging centre to study schizophrenia and other mental illnesses; it had been established by New Mexico's five-term US senator Pete Domenici.

During the trip, Dr Clark wined and dined us. One night towards the end of our visit, after he had impressed us with all the resources and opportunities at the Mind and UNM, he put forth his sales pitch. Dr Clark wanted to know if Vince and I would move to New Mexico.

Vince and I were both flattered, but we told Dr Clark that we were happy in Connecticut.

But Dr Clark persisted. He told us that it would be a joint recruitment by the University of New Mexico and the nonprofit Mind Institute. The university would give us our faculty lines, and the Mind would house our research laboratories and provide our start-up packages.

Dr Clark was a good negotiator. He had already got me to confess my frustrations about working with psychopaths in the community in Connecticut. Moreover, the Connecticut Department of Corrections had been unwilling to transfer inmates from prison to the IOL for research studies. Dr Clark knew I was interested in alternative solutions.

So I told him that I always dreamed of having a mobile MRI and taking it into prisons. And then Vince volunteered his 'say no' offer – a high-field MRI for large-scale research studies. We told Dr Clark that if he could make our requests happen, we would become New Mexicans. Vince and I didn't think there was any chance Dr Clark would find the resources to meet our requests. We were wrong.

---

*   The Mind Institute is now named the Mind Research Network.

Just a week later, Dr Clark called and said that I could have my mobile MRI and Vince could have a new high-field MRI for his research. The Mind Institute would also create a huge database to house all the brain imaging data. They called it a *neuroinformatics system*. That was Vince's baby.

Vince and I looked at each other across the conference room table and stared down at the speakerphone in disbelief.

'Could you please repeat that, Dr Clark?' I asked.

There was a laugh on the other end of the line. Then Dr Clark said, 'We want you here. And we believe your science will have a huge impact on society. You guys think about it and let me know.'

As the call ended, Vince turned to me and said that he wanted to go. I told him I had to first work out if I could get access to the New Mexico prisons. So I booked a flight back to New Mexico.

New Mexico is a large state in terms of square miles but has a population of barely more than two million people. Because of its small population, politicians in New Mexico are rather accessible. Only a few weeks after Dr Clark's call, I was able to get an appointment with New Mexico governor Bill Richardson's director of cabinet affairs. Apparently, Senator Pete Domenici, the founder of the Mind Institute, had made a phone call and a meeting was quickly arranged.

The governor's legal team had already reviewed and verified that my research was legitimate. My work was funded by NIH, and it had been approved by the ethics boards at Hartford Hospital, Yale University, and the US Office of Human Research Protections.

The director of cabinet affairs said the State of New Mexico welcomed research in general, and in particular any research that was targeted to reduce the incidence of antisocial behaviour and substance abuse. Next, I was sent over to meet the New Mexico secretary of corrections.

Secretary of Corrections Joe Williams was a large man. Over six feet tall and barrel chested, he had a very commanding presence. He shook my hand and said to me, 'The governor called and said we are going to get into the research business, eh?'

'Yes, sir,' I replied. 'And my laboratory is at your disposal if you need any research or have any questions about recidivism, risk assessment, or anything else that we might be able to help with.' I was willing to clean the toilets if I could get back in prison.

Secretary Williams had already picked out the first prison where we could start our research programme, the Western New Mexico Correctional Facility in Grants, New Mexico, about eighty miles due west of Albuquerque. The warden had been briefed and was ready to meet me.

It was amazing what could be accomplished with a single phone call in a small state with a small government. Of course, it helps when an introduction comes from a US senator.

The next morning I drove out to the Western New Mexico Correctional Facility in my best suit and met with the warden. I told the warden that we needed research offices and a place to park a mobile MRI. The warden handed me over to the deputy warden, Deanna Hoisington, a petite blond woman with a quick smile and ready wit.

Deanna and I were joined by Dominic, the head of facilities, who led us over to the medical wing. Dominic had identified four offices they were going to empty out for my team. He would order new furniture to be installed.

The warden had insisted that he assign a correctional officer to the research area in case there were any problems. The correctional officer would be just down the corridor but was within earshot if someone had to shout for help. I knew that my staff would appreciate the gesture; none of them had spent any time in prison.

As we approached the rear door of the medical wing, Dominic pulled out a large brass key to unlock the door. It was nearly identical to the key I had used back in the prison in Canada. We exited the medical wing and stepped out into the bright sunshine; the door led to a fenced-in area just inside the perimeter fencing. Dominic pointed to the sally port and said we would bring the mobile MRI in through that entrance and then around the back of the medical wing and park it here on a concrete pad that he would have built.

The mobile MRI needs both a concrete pad to rest on and an electrical connection. The electrical connection is made through an umbilical cable that drops out of the belly compartment of the mobile MRI trailer. I showed Dominic schematics of the umbilical cable and the concrete pad dimensions.

As we finished our tour, Dominic asked if he could get a picture of his brain when the mobile MRI arrived. 'Of course,' I told him. 'I'm happy to take a quick peek for you.' Over lunch Dominic confessed that he too had played football in high school; he wanted to make sure everything was okay.

The last component of my recruitment was interviewing with the faculty at the Department of Psychology at the University of New Mexico. I prepped a new lecture to highlight the development of my paralimbic dysfunction model of psychopathy. I agonized over the tempo and tone of the lecture. I started with the problems psychopaths pose for society, then focused on the latest science, and then on the potential impact this research might have on society. I called it my hourglass talk – it started big, got narrow, then finished with the big-picture implications for society. The psychology department voted to offer me a faculty position.

At the same time, Vince interviewed in the electrical engineering department at UNM. As I remember, he too got a unanimous vote from the faculty to hire him.

Vince and I then got together and reviewed our offers. We had been working together for nearly a decade and had developed a close friendship. The Mind Institute and the University of New Mexico had exceeded our expectations. Our *say no* offers had been met. Vince extended his hand and I shook it firmly.

'I'm all in,' he told me.

'I call,' I answered.

Neither of us was looking forward to the next step, the onerous job of telling our close friends and colleagues in Connecticut that we were leaving.

My meeting with Hank was difficult. Hank had become a surrogate father to me, since my own father had passed away while I was at university. Hank and I had climbed mountains together, literally. We had climbed Mount Whitney together one year, Mount Shasta the next, and we had become extremely close. I knew my recruitment would hit him hard. But in the end I was less emotionally prepared than he was.

As I explained the offer I had received to Hank, I broke down.

Ever the consummate professional, Hank told me that he understood my decision. In fact, he insisted that I *had* to take the offer. He told me to knock off the sadness and let's celebrate this new chapter in my life. And who knows, he added, perhaps some day I would return to Connecticut.

I had made my decision. Now I only hoped I could partially duplicate some day the kind of thriving research environment that Hank had created at the IOL.

Prison Efficiency

Three days after I returned to Connecticut, Dominic, the head facility director at Western New Mexico Correctional Facility, called to tell me that the mobile MRI pad and electrical connection were ready to go.

'I had the inmates build the concrete pad and dig the road along the back of the medical wing. The electrician was able to trench out the connection. So we are all set.'

Dominic must have missed the conversation with the warden about the fact that the mobile MRI hadn't been designed or built yet. Dominic was unfazed when I explained; he just said he wanted to be first in the queue for an MRI once the system was deployed.

Next, I contacted Siemens, GE, and Phillips. I had the sales reps all arrive at the same time to discuss the mobile MRI options for their equipment.

Siemens promised to put the fastest MRI scanner in their product line in a mobile trailer. Siemens had never put their superfast scanner in a trailer before. We worked out a deal that included the most rigorous compliance testing possible. Siemens guaranteed that the system would meet all my specifications — or it would be free. I was impressed. They had a lot of confidence in their engineering. It was one of the simplest decisions I would ever make. Siemens was a go.

Siemens directed me to the three companies that make mobile MRI trailers in North America. I flew to Chicago and then to Los Angeles to interview with two company executives who were willing to develop the world's first mobile functional MRI system. My final stop was Medical Coaches Inc, n the little town of Oneonta, New York.

I drove up to Medical Coaches with a couple of my postdoctoral fellows from New Haven. We met Chief Operations Officer Len Marsh and the vice president of engineering, Dick Mattice, at the Brooks House of Bar-B-Qs for lunch. Marsh and Mattice spent the lunch describing the manufacturing process they followed to meet Siemens's specifications. They exclusively built for Siemens to the German standards for the MRI trailer. Medical Coaches was the best.

I provided Medical Coaches with my sketches, and we agreed to apply for a patent together on the changes we would design into the trailer. We broke down the trailer from the ground up and reconstituted it with special features that would enable installation of the special hardware required for high-speed functional brain imaging. I incorporated changes based on all my experiences designing two other MRI rooms and added all the bells and whistles a researcher could ever need.

When we chose the location of the emergency quench button, I could not help but notice that the large red button resembled the emergency call button from the prisons in Vancouver.

Dick Mattice wanted to install the button in the middle of the MRI room so that everyone could readily find it.

We needed the emergency quench button in case there was an accident where someone was in danger of being hurt while in the scanner, such as if a metal object in the room flew into the magnet. The magnetic field of the MRI is always on; this button would effectively turn it off. It's a last-resort safety measure. Pressing the quench button would collapse the superconductive magnet. But pressing the button can ruin a $2,000,000 magnet. And even if it does not ruin the MRI's magnetic field, it typically costs about $100,000 just to replace all the liquid helium that is vented. The button would have been irresistible to Shock Richie.

I explained to Mattice that we would be scanning inmates like Shock Richie, and that putting the button there was just asking for it

chapter 8

Teenage 'Psychopaths'

> FACT: Incarceration in a US maximum-security juvenile prison can cost $514,000[1] per year per youth.[2]

Any parent will tell you that the teenage years are par-ticularly challenging. Scientists have shown that disruptive and antisocial behaviour peaks in adolescence.[3] Indeed, getting in trouble during this period of development is normal. The majority of boys and a significant number of girls reported engaging in some sort of antisocial behaviour during their teenage years. Fortunately, most delinquent behaviour is limited to relatively minor acts, such as vandalism, alcohol and minor drug use, occasional fights, risky car driving behaviour, and other mischief. However, a minority of youth have a trajectory of antisocial behaviour that escalates into more severe, even violent, antisocial acts.

When we left Brian and Eric, both were thirteen years old and had just been sentenced to juvenile detention for multiple criminal offences (see Chapter 6). Brian was convicted of burglary charges, and Eric was convicted of drug possession and intent to distribute, as well as numerous burglaries. Brian was sentenced in Illinois; Eric was incarcerated just across the border in Wisconsin.

Brian

Following a short stint in juvenile detention at age thirteen, Brian
was placed in a boys' group home for several months before transitioning back to his parents. Upon returning home, he tried to have
anal sex with one of his younger brothers. Brian's parents, shocked,
believe he learned the behaviour from other boys at the group home.
However, they did not seek psychiatric or psychological treatment
for Brian. His aberrant sexual behaviour escalated in the next year
when Brian coaxed a ten-year-old girl at a train station to sneak off
and have sex with him. When Brian and the girl returned to the
train station, the girl told her parents what happened. The girl's parents took Brian home and discussed the incident with his parents.
But both sets of parents decided to keep quiet about the event.

Brian continued his promiscuity, having sex with multiple partners. At one point a teenage girl came to the house with a baby claiming that it was Brian's. The (grand)parents on both sides agreed the
baby should be given up for adoption. During this period, Brian also
had a sexual relationship with a much older married woman. He
was arrested for stealing a watch that he planned to give to her.

As Brian started high school, he initially received above-average
grades. However, his schoolwork declined after a couple semesters.
He started drinking alcohol and smoking marijuana daily. The
marijuana, he later said, helped to alleviate the headaches that continued to plague him.

Brian's home life was chaotic. Brian's father's drinking problems
continued and he lost his job again. The family was evicted from
their home again for failing to pay the rent. Brian's parents' fighting escalated and led to a domestic battery charge against James,
the father. Neighbours reported that the children had little to no
adult supervision. Genevieve, Brian's mother, became reclusive from
neighbours and friends. She too had a drinking problem.

Brian's pranks included pointing a realistic cap gun at little children and watching as they ran for cover. The children told their
parents, and there was a panic that there might be a shooter in the

neighbourhood. Brian was arrested by police at gunpoint and eventually charged with disorderly conduct.

Brian continued to commit burglaries, sometimes breaking into the same homes he had been caught burglarizing previously. He did not plan well, and he was caught with the stolen property a number of times. Judges' sentences ranged from probation to several months in juvenile detention or boys' homes, from which Brian would run away.

Other boys sexually assaulted Brian at these homes, and Brian was treated for anal warts and haemorrhoids.

When released, Brian was always sent home, where he quickly returned to burglarizing homes and businesses. One winter he was caught robbing a Kentucky Fried Chicken and sentenced to several months in juvenile detention.

Brian reported to a boys' home supervisor that he was hitchhiking and was picked up by a man driving a large pickup truck with construction material in it. The man drove to a secluded location and forced Brian to perform oral sex on him. He then dropped Brian off at a petrol station and drove off. Brian went in and bought a can of soda to rinse the ejaculate out of his mouth. Weeks later Brian would identify the man to the boys' home supervisor from a picture in the newspaper when the man was arrested for similar crimes.

The family chaos and Brian's criminal behaviour led to a cycle of short-term incarceration and placement in various boys' homes. His drug use and other behavioural problems escalated.

Brian attended three different high schools in as many years. He often attended school on day passes while incarcerated, returning by curfew each night. He did not participate in any school activities or group sports. Eventually, he dropped out of high school completely.

During his stints in group homes or juvenile detention, psychologists reported that Brian was quiet and did not get involved in any group activities or organizations. Some described him as sneaky and aggressive and stated that he had an oppositional and defiant attitude towards adults and authority figures. Brian did not make friends easily, and he preferred to be alone. Subsequent counsellor interviews reported that Brian had 'poor affect' and judgement and was irresponsible and goalless. Nevertheless, his IQ tested at or above 122, placing him in the superior range.

At age seventeen Brian was released again into the custody of his parents. Within a few weeks he had a serious fight with this mother, who kicked him out of the home. On the streets, Brian survived by committing burglaries for money. He moved in with an older girl for several months. The relationship soured and he went on the road, vagabonding from state to state. He was arrested in Iowa for a burglary and served several months in juvenile detention. But three meals a day and a warm bed were a welcome change from living in his car.

Upon his return to Illinois, Brian was arrested on suspicion of burglarizing a store. Just before his eighteenth birthday, Brian was arrested again and charged with resisting arrest, aggravated battery, criminal damage to property, and unlawful use of an intoxicating compound. Brian got in a fight with police on the way to lock-up and was treated at a local hospital for a broken nose. Released just after his eighteenth birthday, he travelled to Arizona and California for several months, eventually returning to Illinois, where he was arrested for burglary and possession of marijuana. Between the ages of thirteen and eighteen, Brian had spent more time incarcerated than on the streets.

Eric

By the age of thirteen, Eric had been charged with thirteen separate offenses, including possession and trafficking of marijuana, two counts of battery, two counts of arson, weapons charges, and several car thefts. He was on probation when he was caught in a car he had stolen. One week later, he got into a fight with another boy at school and pulled out a gun. While those charges were being processed, he was placed in a foster home; he immediately ran away. Once apprehended, he was moved to a more secure group home. He ran away from there five times in the first month and became well known to the police. In his fourth month at the secure group home, he attacked another boy during an argument. When he learned he would be charged with assault and battery, he set fire to the group home. Around the time of his fifteenth birthday, he was sentenced for three years to a maximum-security state corrections institution for boys – a sentence reserved for only the most severely disruptive youth. The prosecutor argued his violent temperament, prolific use of weapons, especially handguns, and inability to learn from experience put Eric on a trajectory to commit violent crime and claimed he must be incarcerated for the safety of the public.

Eric's behaviour was no better in juvenile prison. He argued and swore at staff, made sexually crude comments to female staff, got into fights with other youth, and destroyed property. Over the course of his first year in the institution, his situation deteriorated steadily. In one two-month period he accumulated forty-nine institutional infractions, including two for manufacturing weapons and two for battery. (The penalty for these infractions was typically to spend several days in 'security' isolation.) Staff reports described him as having a 'very explosive temper'. The staff took several steps in an effort to control his aggressive behaviour, but all met with more aggression. He was eventually placed in an isolation cell for twenty-three hours a day. Even in this setting, he threatened and swore at staff and pounded on the steel door of his room for hours at a time. During his hour outside his cell, he would repeatedly threaten staff, destroy common area property, and refuse to re-enter his cell. After three months in an isolation cell, he was transferred to another, smaller, 'supermax' facility designed to manage the unmanageable.

When he arrived at the supermax facility, he was calm and co-operative according to the initial assessments. During the intake interview, he seemed to enjoy being the centre of attention. He explained that none of the things he had done were that bad and that none were felonies (his record actually showed six felony charges), and he blamed the two boys who were with him when he stole his last car. He admitted that he knew he was not living the way he should, that he should listen to his mother and go to school, but he found that too boring. He claimed that the first time he was in lockup was the worst time of his life. Still, he said that he could not ever recall feeling depressed for more than a few hours. Although he knew his life was not going well at the moment, he was not worried about his future. He didn't consider himself to have failed at anything in his life. 'I'm too young to have failures,' he said. He claimed he was almost always happy, and he rated his self-esteem as a 10 (highest) out of 10. One psychologist noted his arrogance and narcissism were pathological.

On his admission testing to the supermax facility, Eric was scored as being free of emotional distress. Yet he endorsed questions that reflected the attitude that others could not be trusted and generally held malicious intentions towards him. He agreed with questions that described violence as a necessary part of life and agreed that a person had to use violence from time to time to keep others from taking advantage of him.

When asked about school, he admitted that he had been suspended or expelled many times for disrupting class or selling drugs or cigarettes, but mostly for fighting. Eric estimated that he got into a fight at least once a week in school. He said the fights were always provoked by other kids who 'got in my face'. But he later admitted that he usually threw the first punch. Eric was several years behind in school, but he had average intellectual abilities and no learning difficulties. He said that he had spent most of his time in 'security' in the previous institution he was confined to because he kept 'blowing up and cussing at staff'. He attributed these problems to 'staff antagonizing me'.

The first evening at the supermax facility he complained about the food and then destroyed all the items in his cell. When allowed a phone call with his mother, he pulled the cord out of the wall after she told him to stop swearing at her. His temper was rated as one of the most explosive in the history of the facility. Eric's level of aggression even surprised the experienced staff.

Eric particularly disliked being told what to do. He continued to get institutional infractions for disrupting groups or destroying property. He would spend the remainder of his adolescence in the supermax facility. He was due to be released just after his eighteenth birthday.

The supermax psychological assessment team completed a battery of tests on Eric. One of the tests, the Hare Psychopathy Checklist-Youth Version, was used to help understand Eric's risk for recidivism and treatment amenability. Two psychologists independently rated Eric as having a score of 34 out of 40 possible points, placing Eric high in the test range.

The 'Psychopathic' Teenager?

The Psychopathy Checklist is a potent predictor of recidivism in populations of incarcerated adult males. Inmates who score high on the Psychopathy Checklist are four to eight times more likely to commit new crimes upon release than inmates who score in the low range. Nearly all adults with psychopathy that I have interviewed had criminal charges and convictions as teenagers. And those who didn't have criminal charges as teenagers readily admit to significant interpersonal conflicts, and impulsive and poorly planned activities, often of a severe antisocial nature.

It is these findings that have led a steady stream of researchers to develop procedures for early identification of those teenagers on a trajectory towards psychopathy. One goal of this work is to identify high-risk youth so that they can be targeted for interventions.

The most significant attempt to develop a valid and reliable clinical assessment of psychopathic traits in youth and teen years has been the Hare Psychopathy Checklist-Youth Version.[4] The Youth Psychopathy Checklist is based on the original adult version of the Psychopathy Checklist. Like its adult counterpart, the Youth Psychopathy Checklist has 20 items, each of which is scored on a three-point scale (0 if the item does not apply to the individual, 1 if the item applies somewhat, and 2 if the item definitely applies to the individual). The resulting scores range from 0 to 40, and, as in adults, the cut-off of 30 is used to indicate high levels of the traits.

But the academic field tries not to label teenagers as 'psychopaths'. The word *psychopath* is a heavily loaded pejorative, and researchers do not want to prejudice a teenager with such a label. Instead, the affective symptoms of psychopathy in youth and teenagers are referred to as *callous and unemotional traits*. Teenagers who score high on the Youth Psychopathy Checklist are commonly referred to as 'callous conduct disordered youth'. This label is designed to be less stigmatizing than the term *psychopath*. The traits and behaviours that constitute callous conduct disorder as assessed by using the Youth Psychopathy Checklist are listed in Box 4.

BOX 4

Items of the YOUTH Psychopathy Checklist

1. Impression Management
2. Grandiose Sense of Self-Worth
3. Stimulation Seeking
4. Pathological Lying
5. Manipulation for Personal Gain
6. Lack of Remorse
7. Shallow Affect
8. Callous/Lack of Empathy
9. Parasitic Orientation
10. Poor Anger Control
11. Impersonal Sexual Behaviour
12. Early Behavioural Problems
13. Lacks Goals
14. Impulsivity
15. Irresponsibility
16. Failure to Accept Responsibility
17. Unstable Interpersonal Relationships
18. Serious Criminal Behaviour
19. Serious Violations of Conditional Release
20. Criminal Versatility

A considerable amount of research suggests that the affective and interpersonal traits of psychopathy are relatively stable from adolescence to adulthood. The best evidence of this comes from longitudinal studies of high-risk youth.[5] These studies find that, without intervention, CU traits remain relatively unchanged from teen years through early adulthood.

Studies have also shown that Youth Psychopathy Checklist scores accurately predict future violent and antisocial behaviour.[6] The Youth Psychopathy Checklist has been proven, over and over again, to show better predictive utility than other risk factors for violent outcomes. Youth with high CU traits are more violent, begin offending earlier, and have a greater number of police encounters than low-risk peers.[7] Indeed, within the legal system over the last decade, the single biggest increase in the use of psychopathy evidence in youth proceedings has been in violent risk assessment.[8]

Forensic clinicians, prosecutors, and ultimately judges are increasingly called to make predictions of future adolescent violent behaviour. Given that research shows that CU traits predict future violence, it seems prudent that these traits be accurately assessed in high-risk youth.

However, decision makers in the youth forensic arena must continue to follow the best measures for quantifying CU traits, and be aware of the strengths and limitations of the instruments and their predictive utility for the context in which they are being applied. Youth Psychopathy Checklist scores do not accurately predict future behaviour in every context and for every type of youth offender. For example, in teenage girls there is not currently enough peer-reviewed evidence that Youth Psychopathy Checklist scores reliably predict future violence.

These traits are not based upon a single act or a single area of someone's life. The trait must be present in the majority of the individual's life — at home, work, and school and with family, friends, and neighbours. It is only when the trait typifies the majority of someone's life that we score the item high. Indeed, when working with a teenager who has committed a crime, psychologists are asked to ignore the index offence when scoring the Youth Psychopathy Checklist. The assessor should get nearly the same score without taking into account the index crime. Ignoring the index offence will ensure that one serious crime does not create a halo that impacts the scoring. Psychologists are assessing traits, not just individual bad acts. A case in point is that of Chris Gribble.

New Hampshire Tragedy

Chris Gribble was a nineteen-year-old first-time offender who participated in one of the worst crimes in New Hampshire history. Chris and three other boys (Steven Spader, William Marks, and Quinn Glover) broke into a home in the middle of the night. Spader, the mastermind, had convinced the three other boys to target a rural home, break in and rob the place, and kill anyone they encountered. It was a quest to commit a thrill killing.

On the fateful predawn morning of 4 October 2009, the four boys broke in through the basement window of the Cates family home in rural Mont Vernon, New Hampshire.

As the two other boys searched the house for things to steal, Spader and Gribble made their way to the master bedroom where forty-two-year-old Kimberly Cates and her eleven-year-old daughter, Jamie, were sleeping. David Cates, the husband and father, was away on business.

Spader viciously attacked Kimberly with a machete while Gribble used a knife to repeatedly stab Jamie. Kimberly succumbed to her wounds, but Jamie miraculously survived by playing dead. Despite more than a dozen serious wounds, Jamie managed to call emergency services and make her way to the front door, where the first responder, veteran police sergeant Kevin Furlong, applied first aid and called for an ambulance.

The horrific, senseless crime shocked the community.

The four boys were quickly arrested the following day after tips poured in to police.

Donna Brown, a seasoned public defender, was assigned to Gribble's case. Donna was familiar with my research and called me to consult on the case. She felt that something was wrong with Chris, but she was not sure what, if any, illness he might be suffering from. Donna also wondered if we should get an MRI of his brain to rule out any clinical abnormality, like a tumour.

Gribble came from a very unusual background. He had been home-schooled by his mother since childhood at their rural home.

He had very few social experiences as a child or teenager as he rarely left the home.

The Gribble family was devoutly Mormon and Chris's limited social experiences occurred in that venue. As a teen he dated a girl for two years whom he met at church. The two never kissed, but rather held hands while they walked the church grounds on Sundays. The girl would describe Chris as very socially awkward.

Chris used no drugs. He tried to drive his car fast a couple of times but got scared and slowed down. There was no history of his lying or being manipulative.

In many areas of his life, his emotions were in the normal range but were a bit stilted — he was more immature than shallow and more poorly developed than callous. There was no evidence that he had an inability to feel guilt for petty transgressions. Indeed, he had a normal relationship with his brother and a few other teen friends.

There was no evidence of his leeching or living off others. Chris had a job for many months working at the sandwich shop Subway. He did not have a history of stealing from his parents, friends, girl-friends, neighbours, or employers. He did not excessively borrow money from family or friends. He did not try to live a life beyond his means.

Chris's first sexual experience did not occur until he was nineteen.

There was no evidence of any severe behavioural problems as a child or teen and no encounters with the police before the index offence. He was not impulsive or irresponsible.

One area of his life that was particularly distressing to Chris was his relationship with his mother. Chris felt that his mother had abused him, and he was sent to a psychologist as a teenager to get help.

During our interview, I asked Chris to describe the abuse he claimed he'd suffered. He related stories of being spanked as a child for not doing his homework. He said that his mother would put him on 'restriction' for not completing his chores or other duties.

I pointed out to Chris that the vast majority of stories he told about his mother would not constitute child abuse.

In a rare moment of clarity, Chris volunteered that he recognized that many people would not think that his mother's behaviour was abusive, but it was a source of great distress for him. His *perception* of her behaviour led him to interpret it as abusive.

Chris was in complete denial about the crime he had committed. Moreover, instead of accepting that he made a mistake, Chris tried to glorify the crime.

Then Chris told me that he was a psychopath.

His statement took a minute to settle in. I've never had an individual tell me he was a psychopath with such enthusiasm. To me, it was a signal something was amiss.

I asked Chris if he knew what a psychopath was. He said that a psychologist who had treated him for his problems with his mother had told him he scored high on the self-report MMPI psychopathic deviate subscale and told him he was a psychopath.

'A psychologist told you were a psychopath during a treatment session based on an MMPI profile?' I asked him.

'Yes,' he said.

I was so shocked that it took me a minute to compose myself and continue the interview. Not only had this psychologist committed an enormous ethical error by telling a teenager he was a psychopath, but he also based his diagnosis on a self-report test that has dubious utility to assess the condition.

The impact of the psychologist's diagnosis on Chris was unmistakable.

Armed with this new label, Chris had decided to act like a psychopath. It had become a sort of a self-fulfilling prophecy.

Psychologists are concerned that if people start to believe something to be true (even if there is no evidence of it), they may precipitate the actions they are trying to avoid. For example, a woman might falsely believe her marriage is failing and change her behaviour in such a way that actually causes more marital problems and eventually leads the marriage to fall apart. This process has been termed a *self-fulfilling prophecy.*

Prior to meeting Chris, I was unaware of anyone being told he was a psychopath and then having that suggestion precipitate psychopathic behaviour. But that is exactly what happened to Chris. Before the crime, there was very little evidence of psychopathic symptoms in his background. But after the crime, Chris wanted the world to believe he was a psychopath.

However, Chris did not present like a psychopath; he was just developmentally disabled in terms of his social interactions and ability to emotionally connect to people. In addition, I noticed that there was something odd about Chris.

My clinical 'spidey sense' was going off in a big way.

Chris's eye movements were slightly abnormal, but not in the way that psychopaths' eyes are abnormal. His eyes were completely disconnected from thought and affect. I thought his abnormal eye movements were a reflection of brain disturbances due to his aberrant social and emotional development.

Underlying Chris's thoughts and expressions was an incredible emotional and intellectual immaturity. He had a very poorly developed worldview, and his social skills were dismal. His IQ was also well below normal.

'Once we had broken into the house, we prevented the alarm from going off by walking slowly so as to not set off the motion sensors,' Chris explained. He then proceeded to stand up and demonstrate how slow he could walk. He shuffled his feet and inched across the linoleum floor of the prison interview room.

Public defender Donna Brown was shaking her head with a look of complete embarrassment on her face.

'I was watching really closely and would slow down even more if the motion sensor turned from green to red. But I was able to go slow enough that it didn't detect me.'

Chris proceeded to inch his way around the Cates home until he found the electrical circuit breaker panel. He turned off all the power to the house in an effort to defeat the house burglar alarm.

I glanced over at Donna; she had her head down and she was partially covering her face with her hand to keep Chris from seeing her expressions.

I informed Chris that home alarms in general and motion sensors in particular have a battery backup to prevent criminals from being able to defeat them by turning off the power to the house. Homes often lose power in storms; the battery back-up protects the alarm system.

I also told him that when a motion sensor turns from green to red, it means it has detected motion.

The reason the alarm did not go off, I told him, was because the Cateses had not armed it that night.

'Oh,' he muttered. 'I guess you learn something new every day.'

There were countless other examples that indicated Chris was a very unsophisticated individual.

For example, Chris was extremely susceptible to suggestion. He would watch television shows and then believe that he could do what he saw on TV. He thought that by watching enough *Law and Order* episodes he could become a lawyer.

I came to understand that Chris's belief that he was a psychopath was akin to a delusion.

Let me explain. I once had to perform a risk assessment on a prisoner who believed that his father was a famous Hollywood actor. 'Michael', as I will call him, had developed this belief about his parental lineage and had in fact driven long distances from his home in Canada to Los Angeles to see his famous father. He had broken into a house he believed to be owned by the actor and started living there. After he was arrested, he continued to believe, despite ample evidence to the contrary, that his father was the famous actor. Michael's real father used to visit him in prison; Michael thought of him more as a friend than a father. Michael's delusion lasted his entire adult life.

I met Michael some twenty years after his third break-in of the Hollywood actor's home. The third crime had triggered special provisions in Canada that placed Michael in the 'dangerous offender' category, and he had received a long prison sentence.

The treatment team had been trying to break Michael's delusion for years. In the past year, Michael had finally started telling the treatment team that he no longer believed that actor was his father.

I had been asked to evaluate whether Michael was fit for parole.

Two hours into the interview we had a good rapport going, so I subtly sprang the question on Michael I had most wanted to ask, inquiring what his mother and father did for a living.

Michael took the bait. He told me that his father was an actor and began to tell me stories about his famous father. As he spoke, he looked right into the video camera recording the risk assessment interview. He was just telling the truth — as he believed it.

I showed the video to his treatment team. They looked confused. 'When did you do the interview?' his main therapist asked.

'Today,' I replied. They looked at each other in disbelief.

It turned out that Michael had grown tired of people telling him that the actor was not his father, so he told them what they wanted to hear. But he still believed it to be true.

I recommended that he not be released from prison. I knew that he would drive straight down to Los Angeles and break into the actor's house again. I thought I would save the actor that hassle.

Michael was not violent; he was not a problem at the prison. He just suffered from a delusion, and his treatment team had no idea how to fix it. None of us did. We only knew that he had to be contained in order to prevent him from breaking the law.

I believed Chris Gribble suffered from the same type of delusion. Chris's delusion was that he thought he was a psychopath. And that made Chris very dangerous.

Chris's psychopathic symptoms are more a result of a break from reality than a reflection of his underlying personality. There was just no evidence from the other domains of his life that Chris has severe psychopathic symptomology. He does not score high on the Youth Version of the Psychopathy Checklist. Indeed, he scores below average for a youth in prison.

Chris blames his mother for his problems and claims he has wanted to kill her for years. However, he never actually tried to hurt his mother. A child or teen with elevated CU traits would typically not hesitate to hurt or lash out at his or her parents. Indeed, many youth with elevated CU traits fight with their parents as easily as they fight with other children on the playground. I have interviewed more than a few psychopaths who had murdered their parents.

The fact that Chris did not act on his thoughts of hurting his mother confirms that he had some internal controls.

But those internal controls had been completely overcome by the social pressures exerted on him by Steven Spader. Spader had a long history of criminal activity and encounters with the police. He abused alcohol and drugs. He had many different girlfriends and frequent infidelities. He stole from family and friends. He manipulated others into dealing drugs for him. He fit the profile of a psychopath far more than Chris did, although I never tested Spader personally.

Chris was in the wrong place at the wrong time. He had joined the wrong peer group. Had he received the right help, that terrible crime might never have taken place.

Epidemic of Spree Killings

Tragic crimes like those committed by Gribble are unfortunately far too common. It seems that every few weeks there is a mass killing committed by some disturbed young person. Eric Harris and Dylan Klebold at Columbine High School; Seung-Hui Cho at Virginia Tech University; Jared Loughner at a political rally in Tucson, Arizona; James Holmes in a cinema in Aurora, Colorado; and Adam Lanza at Sandy Hook Elementary School in Newtown, Connecticut, are just a few examples of the type of heinous spree killings that plague American society.

I am often asked whether the perpetrators of such crimes are psychopaths. Certainly, the act of mass killing can be described as *psychopathic*, but does that clinical label fit the personality of the attacker? For the vast majority of spree killer cases, the answer is no – the offender was not a psychopath. Most killing sprees are committed by individuals who suffer from psychosis, not psychopathy. Recall that psychosis is a fragmentation of the thinking processes in the brain that leads to symptoms like hallucinations and delusions.

Hallucinations come in all modalities – a patient with psychosis might hear voices that are not there (auditory hallucinations), see things that are not real (visual hallucinations), feel things that are not real (touch or somatosensory hallucinations), smell things that are not there (olfactory hallucinations), and taste things that are not real (gustatory hallucinations). All sensory modalities can be affected in a psychotic illness. Hallucinations can take many forms, but the most dangerous types are known as *auditory command* hallucinations. Command hallucinations are voices that the patient hears instructing him or her to perform some behaviour. Sometimes the command hallucinations tell the patient to kill people. The voices that constitute command hallucinations can be powerful forces that compel a person to act in an irrational manner.

Delusions are also a central symptom of psychosis. Delusions come in many forms, most of which are relatively harmless. But when a delusion robs a person of understanding the impact his or her behaviour will have on others, it can be very dangerous. Delusions can override normal rational thinking processes, and some delusions lead people to commit homicide.

It is important to recognize that most patients with psychosis do not harm others. But psychotic symptoms do increase an individual's risk for committing homicide.

Schizophrenia is one of the main disorders associated with psychosis. The prevalence of schizophrenia is estimated to be below 1 percent of the general population, but patients with schizophrenia comprise between 5 and 20 percent of all homicide offenders.[9] A worldwide study found that 1 in 629 patients in his or her first episode of psychosis ended up committing a homicide.[10] In comparison, only 1 in 25,000 individuals in the general population of the United States will commit homicide. (The number is lower in the United Kingdom.) Thus, psychosis is associated with a fortyfold increased risk for homicide than that found in the rest of the population. The average age of onset of psychosis in men is eighteen and in women it is twenty-five, and research shows that the vast majority of homicides committed by such patients occur within the first year of illness onset.

However, if psychotic patients receive treatment, the risk for homicide drops from 1 in 629 to 1 in 9,090.[11] The most effective way to prevent homicide in patients with psychosis is to get them early and effective evidence-based treatment. Unfortunately, as has been well documented in the cases in Arizona, Virginia, and Colorado, despite ample evidence the perpetrators were mentally ill prior to the commission of their crimes, they did not receive good treatment.

In summary, the majority of spree killers who go on rampages suffer from some sort of psychotic illness that includes command hallucinations and delusions. As we have reviewed, personality traits, especially psychopathic personality traits, are based on one's entire life history and in all domains of a person's life. An isolated violent crime, even one that leads to multiple deaths, does not by itself warrant a diagnosis of psychopath. Also, hallucinations and delusions are not symptoms of psychopathy; thus most individuals who commit spree killings do not meet criteria for psychopathy.

chapter 9

Mobile Imaging

FACT: The cost of crime in the united states is $3.2 trillion per year – an amount greater than the expenditures for all health care in the country.[1]

One of the board members of the nonprofit Mind Institute in New Mexico[2] was Dr John Nash, a Nobel Prize–winning economist from Princeton University. Dr Nash's life has been chronicled in the film *A Beautiful Mind*, in which he was played by actor Russell Crowe. Dr Nash was able to fight through his own mental illness to create game theory in economics.

Dr Nash felt the science I was conducting would be a game changer in forensics, and he was one of those who convinced me to accept the job offer from the Mind Institute and the University of New Mexico.

Relocating my laboratory from Connecticut to Albuquerque, New Mexico, was not simple. My laboratory was composed of fifteen full-time research staff, two Yale psychology graduate students, and a dozen undergraduate students working on various projects. Moreover, I had to continue to make progress on my current grants in Connecticut, as well as set up a parallel laboratory in New Mexico so that my staff could transition smoothly.

During long meetings, we finalized plans for the undergraduate and graduate projects to run to completion in Connecticut. I then flew all fifteen staff members out to New Mexico to tour our new lab space at the institute. My staff visited during the world-famous hot air balloon festival in Albuquerque. The city was alive with energy; every morning we saw hundreds of balloons floating around the city, a breathtaking sight.

Upon returning from the visit, my staff organized a meeting and called me in. They had all agreed to the move. Dedicated young professionals, they shared my interest in using the latest neuroscience to better understand psychopaths. I was honoured that they were all willing to relocate.

The New Mexico Corrections Department organized a weeklong training session for my staff on the ins and outs of working in the New Mexico prisons. My senior staff attended the training and developed a manual to train the rest of our staff.

Once the plans for transitioning were finished, I turned my attention to finding help completing my nonforensic projects. I was still actively pursing brain imaging studies on the neural basis of psychosis and how neuroscience could assist in differentiating patients with schizophrenia from those with acute psychotic bipolar illness.

Finally, I travelled up to Oneonta, New York, to Medical Coaches Inc, where our prototype mobile MRI was being built. My laboratory's ability to collect data for my grants depended on the success of this new system. It was a risky move. I was definitely all-in; I was going to make this mobile MRI work or go bust.

Birth of the Mobile fMRI

The engineers at Medical Coaches Inc added dozens of my special features to the trailer to enable us to incorporate all the MRI-compatible equipment needed for functional MRI studies. These included mounting five computers on the control room's walls to support our data collection efforts. One computer was used to drive the visual projection system to deliver pictures or video to the prisoners while they were in the MRI. We connected the video output of the computer to a high-definition projection system housed in a shielded cabinet in the rear of the trailer. The projector would throw the video image to a special screen mounted in the MRI bore so the prisoners could watch all the stimuli and videos we planned to present to them. It was a state-of-the-art, high-definition video system.

Another computer recorded and monitored in real time the timing of the stimuli and responses from the prisoners. When you work with psychopaths, you have to ensure that they are completing the tasks you ask of them in the MRI. Indeed, I also set up a computer to track the eye positions of the inmates while they were in the MRI. We could monitor the precise location of where on the video screen the inmates were looking when we presented them with stimuli. This further ensured they were paying attention and not falling asleep!

Another computer tracked and monitored all the research data and interfaced with the Mind Institute's neuroinformatics system. We were going to be collecting terabytes of brain imaging data and we needed to organize, track, and back up all of it.

Finally, the last computer interfaced with the projection system to present a video of a cartoon fish tank to the prisoners. I had found that it was very helpful to start off the scan sessions with the fish tank display so that the inmates would know where the video screen was located in the MRI chamber and where to focus their attention. The fish tank also served to calm individuals who might be a little claustrophobic in the tight confines of the MRI bore.

The electrical, heating, and cooling systems of the mobile MRI were upgraded to handle the additional computers. Special tubes, called *wave guides*, were installed between the control room and the projector room at the rear of the trailer. Through these wave guides we would install the LumiTouch fibre-optic response devices I had designed with a company in Vancouver, British Columbia. These devices were able to collect responses from the inmates while they were performing the cognitive and emotional tasks in the MRI.

I scheduled visits to New Mexico once a month for the remainder of 2006 and prepated to move my laboratory staff in early 2007. We would have the first four months of 2007 to conduct interviews and testing with the inmates before the mobile MRI would arrive in May.

During my visits, I began to establish relationships with the staff of the New Mexico Corrections Department and the New Mexico

Children, Youth and Families Department. We developed plans to set up research laboratories in the Western Men's Prison and the Women's Prison of New Mexico, both of which were located in the small town of Grants, about eighty miles from Albuquerque. We also started a youth project at the Youth Diagnostic Detection Center, the only maximum-security facility for teenage offenders in the state. Finally, we planned a home base for the mobile MRI crew at the Mind Institute headquarters on the campus of the University of New Mexico in Albuquerque.

I also had three grant projects to set up, two supported by grants from the National Institute of Mental Health (NIMH) and one funded by a grant from the National Institute on Drug Abuse (NIDA).

The first NIMH project was designed to test my paralimbic dysfunction hypothesis of psychopathy in adults. My team and I planned to examine the brain structure and function of psychopathic and non-psychopathic inmates. We designed several emotional and cognitive tasks to probe the function of the paralimbic system.

The NIDA project was the largest and most ambitious of the three. We planned to recruit 150 inmates with substance abuse problems into a twelve-week treatment programme. During the treatment programme, the inmates would receive one of three different types of cognitive behavioural therapy to help them stop using drugs. The mobile MRI would be used to scan the inmates before, during, and after completion of treatment. In this way we could examine how the brain changes with treatment. We could also study whether certain parts of the brain or certain functions of the brain predict who will do well and who will fail in treatment and return to drug use.

The project necessitated recruiting and training over a dozen qualified clinical psychologists to deliver the cognitive behavioural therapy. We also had to track down and interview the inmates after they were released in order to collect data about whether or not they had relapsed to substance abuse.

The second NIMH project was conducted at the maximum-security juvenile prison. We planned to examine whether youth with elevated callous and unemotional traits looked similar to or different from adults with psychopathy. We also planned to carefully assess the environments in which the youth had been raised, collecting data on parenting styles, stress and trauma, and other variables that might be related to the development of CU traits. We also planned to examine the influence of drug abuse in teenagers.

The MRI was installed in the custom trailer and driven out to New Mexico for delivery on 15 May 2007.

On the day it arrived, I was the first to climb in to get a picture of my brain. As the MRI surged to life and started whirring and beeping, I could not believe that my dream of using a mobile MRI to study psychopaths in prison had come true.

We had a week to test the scanner before 'accepting' it and finalizing the last payment in our contract with Siemens. The MRI sequences pushed the technology to the limits. And then we had our first hiccup. We fried a circuit board in an amplifier by pushing the unit too hard. Working with the Siemens engineers, we determined that an upgrade was needed. Some of the hardware in the mobile MRI had been shrunk to fit in the small confines of the trailer.

We had to replace some of the mobile MRI hardware with regular-strength hardware. It was a quick fix, and we were back to testing within a couple days. After that first glitch, the mobile MRI stabilized and performed well. The overall quality of the data generated during our testing was amazing. The mobile MRI was more stable than many MRI systems installed in hospitals. Medical Coaches Inc had done a great job building the trailer, and Siemens had tuned the scanner perfectly. Following the week of testing, we installed all our computers and fMRI equipment. Then our driver picked up the Mind Mobile MRI System and moved it out to Western Prison.

I called Dominic, the facilities director at Western Prison in Grants, New Mexico, to tell him the MRI was on its way, thanking him for his patience and letting him know that he could get that picture of his brain now.

Intersection of Law and Neuroscience

Around the same time as the mobile fMRI was being deployed, I was approached by my good friend Dartmouth College professor Walter Sinnott-Armstrong about whether I might be willing to join a research network on neuroscience and law being directed by Dr Mike Gazzaniga. The John T. and Catherine D. MacArthur Foundation had been asked to fund a think tank of psychologists, neuroscientists, philosophers, law professors, and state and federal judges to examine contemporary issues in neuroscience and law. The research network was tasked with deciding and then funding research that would stimulate growth in this emerging field.

Walter mentioned that initial discussions had centred on whether the neuroscience of psychopathy might have an enormous impact on philosophical thinking regarding legal constructs like criminal responsibility, punishment, and capacity for change.

Membership in the network was a significant time commitment. Members would attend three general network meetings per year; two subnetworks would meet every other month. Finally, small working groups examining hot topics would meet as needed. Over the next three years, I travelled at least once a month to the MacArthur neuroscience and law meetings.

At these monthly meetings, I was tossed into an academic melting pot of some of the best scientific and legal minds in the country. My fellow neuroscientists and I described the methods of our field, their strengths and limitations, and their promise to improve mental health diagnosis and treatment. The law professors described the philosophical underpinnings that the law held most dear, underpinnings that some had argued would be completely transformed by the emerging findings from neuroscience. Judges described the difficult decisions they had to make every day in court, raising questions about whether neuroscience might provide information that would make decision making easier.

There were no easy answers to some of the questions raised, but it was clear that modern neuroscience and the law were colliding at many levels and in numerous domains.

The intellectual environment of the inaugural MacArthur Foundation Research Network on Law and Neuroscience was one of the best academic experiences of my career.

Pump Up the Volume

During graduate school, it had taken our lab five years to organize and transport fifty inmates from prison to scan them on the university MRI. My lab scanned fifty inmates on the mobile fMRI in the first week it was deployed in prison.

By the end of the first year, the mobile fMRI had scanned the brains of over five hundred inmates. In just one year we had collected the world's largest database of brain scans from forensic populations.

I had found my remedy to at least one of the issues raised by the editors of the journal *Science*. I was never going to publish a small sample size study again — especially given the potential societal impact of this work. Indeed, if there was one thing that the scholars on the MacArthur Foundation Research Network on Law and Neuroscience agreed about, it was that the neuroscience of psychopathy was going to have a big impact in the legal system.

The prison staff and inmates of the New Mexico Corrections Department were generously supportive of our research efforts. Our relationship was a true reflection of the power of collaboration between a state entity and an academic research programme. Indeed, there were even collateral security benefits. After the first year of research, one of the deputy wardens told me that institutional infractions had gone down since we started doing research with the prisoners. The prisoners had told the deputy warden that they enjoyed participating, and they stayed out of trouble so they would not get put in segregation, where they would not be able to do research. The New Mexico Corrections Department allowed us to expand our research programme into other prisons around the state. After just a few short years, we were working in more than half of the prisons in New Mexico.

I had to buy a fleet of Toyota Priuses for my staff to drive.

Normally, staff are reimbursed on a per mile rate when they drive personal vehicles for work. I had done a calculation and figured out that my grants would be charged over $50,000 per year if we followed that model. I calculated that with a Prius we could save about $150,000 over the course of the five-year grant.

'Buy two' had been the answer from my NIDA programme officer, Dr Steve Grant, when I asked him if I could buy a Prius to save money. We ended up owning four Toyotas by the end of our second year in New Mexico. My staff put over fifty thousand miles per year on each of the cars driving back and forth to prisons around the state. The Toyotas would soon eclipse a *moon unit*, over 252,000 miles, as my staff drove to the farthest corners of New Mexico to complete our research studies.

Structure of the Psychopathic Brain

My laboratory was a buzz of activity. We collected data for faculty from the University of California, Santa Barbara; Harvard University; Vanderbilt University; University of Southern California; Dartmouth College; Duke University; Princeton University; Stanford University; Arizona State University; and Washington University in St. Louis.

We were writing grants as fast as we could to continue to fund our research studies. The article in *The New Yorker* by John Seabrook that I mentioned in Chapter 2 generated a lot of media attention for the research we were conducting, and a few donations came in following the article to help fund our work.

Hearing about our psychopathy research, postdoctoral fellows and psychologists from around the world sent in applications to work in our lab. I was more than happy to oblige and share my passion for studying psychopaths.

I assigned one of my talented postdocs, Elsa Ermer, to analyse the first year of structural MRI data collected from adult male offenders. I wanted to know if the density of grey matter in the paralimbic system was abnormal in psychopathic inmates compared to nonpsychopathic inmates.

Grey matter is composed of the cells that perform computations in the brain. These thinking areas of the brain are contained within the grey matter. In contrast, white matter refers to the tracts that connect brain grey matter regions together.

Scientists have learned a lot about grey matter from the analyses of structural MRI data. For example, we know that, unfortunately, as one ages grey matter tends to decrease in many areas of the brain. But there are also regions, like those that store memories, that increase in density as age increases. As we store more and more memories, parts of the brain get thicker.

We also know that individuals with higher IQs have more dense grey matter than individuals with low IQ. Patients with mental illnesses, like schizophrenia, have abnormal grey matter in the frontal and temporal lobes of the brain.

I had hypothesized that psychopathy would be associated with grey matter abnormalities in the paralimbic system, the parts of the brain that regulate emotion and impulsivity.

Our first analysis was composed of over three hundred adult male offenders. I asked Elsa to work with our computer programmer to analyse the data. We met and planned out the analyses to make sure to examine any variables that might cloud interpretation of the results. For example, we had to ensure no inmates had any history of serious brain injury or other problems that might obscure the results.

I figured it would take weeks, if not months, to analyse an MRI data set of over three hundred inmates.

One Monday morning I got to the office before 8 a.m., and my programmer and Elsa were already there, hunched over a computer.

'Good morning,' I called to them on the way to my office.

'Stop! Come here!' Elsa exclaimed.

She was pointing at the screen in front of our lab programmer, Prashanth Nyalakanti.

'Look at these results.'

It had never occurred to me that the analyses could be completed in just one weekend.

'How can that possibly be ready?' It normally took about twenty-four hours per inmate to crunch the numbers on grey matter density in all the regions of the brain.

Prashanth said that he had written a script to use all the Mind computers – more than a hundred – over the weekend; the analyses were completed Sunday night.

Within each of the Mind's one hundred computers, there were four to eight processing cores. Prashanth had assigned a brain scan to each processing core. What would normally have taken several months of computer cycles had been completed overnight by dispersing the data analyses over five hundred computer processing cores. It was a brilliant piece of code that also used the processing cores only when other staff were not using them.

I was impressed.

Elsa pointed to the screen. A blue colour map indicated the brain grey matter regions that were negatively correlated with Psychopathy Checklist scores. It was a map that I will never forget. The paralimbic system was bright blue (see Figure 8).

Elsa said, 'Isn't that amazing?'

I was speechless. I was able to muster only a weak nod in agreement.

I grabbed a chair and nudged Prashanth over a bit. It was truly amazing. Psychopaths showed reduced grey matter in the orbital frontal cortex, amygdala, hippocampus, insula, temporal pole, and in the anterior and posterior cingulate. The entire paralimbic circuitry was impaired. I had not even fantasized about such a dramatic result. I was blown away.

It was one of those rare moments in academics when a hypothesis is proved to be tru – a hypothesis that I, and my entire lab, had laboured for years to craft and frame. I told Elsa she would remember this day for the rest of her life. Psychopaths' brains *were* abnormal; we had solid proof.

I printed out copies of the maps and ran downstairs to show my collaborator Vince Calhoun. He asked a few general questions, double-checking my assumptions, and then he asked for a copy of the code that Prashanth had written to parallel process the brain imaging data. He too was impressed, both with Prashanth's ingenious code and with our results.

Elsa led the write-up of the results. They were quickly accepted in the *Journal of Abnormal Psychology* — the top psychological journal in our field.[3]

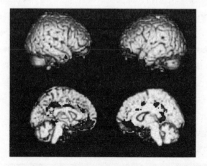

Figure 8. Results from the first structural MRI analyses of criminal psychopaths collected on the mobile MRI system. The shaded regions depict areas of the brain that are atrophied in adult male criminal psychopaths. The areas represent the majority of the paralimbic system of the brain (see note 3).

I presented the brain density results at the MacArthur Foundation Research Network on Law and Neuroscience meeting that month. The law professors and judges stared at the figure for a long time, processing the potential implications. I reminded them that the data were 'correlational', not 'causal'.

I used the following analogy to help the judges better understand the results.

If I put a sling around your arm and tie it to your chest, your bicep muscle will atrophy over time. When you visit a doctor, you will likely be asked if your arm has been atrophied your whole life or whether it was like that because of recent nonuse. The doctor would not be able to tell the difference from just looking at your arm.

The brain too is like a muscle. Brain grey matter density can change over time. Thus, it is not clear from this one data set whether the psychopath's brain was atrophied from birth or whether the psychopath's life experiences had caused the brain to atrophy over time. That is, if you don't use parts of your brain, they can shrivel over time, stop working, and even die.

Again, our results were not 'causal'; that is, we could not tell whether psychopathic behaviour was a direct result of an atrophied brain. But the next analyses I planned to do might help answer this question.

Paralimbic Development

As soon as the adult psychopath grey matter paper was accepted for publication, I set up another meeting with Elsa and Prashanth. I asked them to analyse the brain scans from the two hundred teenage boys from the maximum-security juvenile prison. Most of these youth had been convicted of multiple felonies. Indeed, there was an entire housing unit of youth at the prison who had been convicted of murder. These youth represented one of the most high-risk populations of teenagers in the country.

If the brain scans of the incarcerated youth showed the same results as the adult scans, it would suggest our findings were not likely to be due to atrophy from nonuse. The most parsimonious interpretation would be that the paralimbic atrophy was present from birth. It was a conclusion that followed Occam's razor — it made the fewest assumptions.

It took only twenty-four hours for Prashanth to parallel process the youth MRI data. Elsa added a smiley face to the e-mail subject line when she sent me the PowerPoint file with the data.

Once again I was struck by the results (see Figure 9). The incarcerated youth with elevated callous and unemotional traits as assessed by the Youth Psychopathy Checklist had the same brain abnormalities as the adults with psychopathy. The brain plots of the youth and adults were uncannily similar.

A mixture of emotions ran through me. As a scientist, I felt excited and vindicated that my hypotheses about the psychopathic brain were borne out. As a human being, I felt sad for these youth-

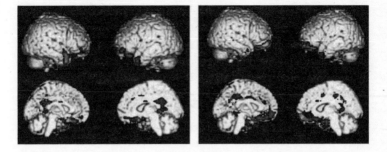

Figure 9. Results from the first structural MRI analyses of incarcerated youth collected on the mobile MRI system (left four images). The shaded regions depict areas of the brain that are atrophied in youth with elevated callous/unemotional traits. The results from adult males (right four images) are shown for comparison. The results from the two groups are strikingly similar. In both groups the majority of the paralimbic system is showing reduced grey matter density or atrophy. Youth data is from Ermer, E., et al. (2013). Aberrant paralimbic gray matter in incarcerated male adolescents with psychopathic traits. *Journal of the American Academy of Child and Adolescent Psychiatry*, 52, 94–103. Adult data is from Ermer, E., et al. (2012). Aberrant paralimbic gray matter in criminal psychopathy, *Journal of Abnormal Psychology*, 121(3), 649–658

with these abnormal brains. It was as if my lab had discovered a new disorder, but we didn't have a cure.

As the mixture of emotions in me subsided, one thing became clear — these results were going to spark a lot of discussion and debate. Debates would occur between academics, between lawyers and judges, and between public officials and the media.

But the results were sound. We could not be criticized for publishing small sample sizes.[4] Indeed, our youth study had ten times as many participants as the closest competitor.

I hoped that other scientists would replicate our findings someday. But for now I was pleased with my laboratory's efforts.

chapter 10

The Decompression Chamber

> Fact: some forms of group therapy may make psychopaths *more* likely to commit new crimes following release from prison than no treatment at all.[1]

Brian

At the age of eighteen, Brian began to use false identifi-cation papers and other aliases to escape criminal charges or get reduced jail time. He was a prolific burglar, sometimes hitting more than a dozen homes a day. He lived in the rented home of his older sister and her family.

Brian did not get along well with his sister, and the two had numerous loud verbal altercations; at one point their altercations led neighbours to call the police.

During the burglary of a school, Brian set fire to the building. He was caught nearby by police, to whom he gave a fake name. The police determined his true identity and charged him with arson. He served less than a year in jail before being released on parole.

Brian returned to burglarizing businesses and homes within hours of being released. As he turned twenty-one, he was arrested for burglarizing a church and was sentenced to two to six years in the state prison.

Prison records indicate Brian's IQ tested well above average. He completed his high school equivalency degree while in lock-up and started taking university-level courses.

Brian made few friends and many enemies in prison. An investigation by the Federal Bureau of Investigation into prison rape concluded Brian was sexually assaulted by other inmates. Brian testified against the other inmates in return for placement in protective custody, where he was kept isolated from the general population.

Brian's co-operation with the authorities and good behaviour while in protective custody earned him early release from prison. But once on the street, Brian quickly returned to a life of crime. At the age of twenty-five, he was arrested on suspicion of sexual assault of a twenty-two-year-old woman. He posted bond and eventually beat the charge when his brother Steve provided an alibi for him on the night of the crime.

Next, Brian hit the road, with brief stays in Texas, Arkansas, and Arizona. He moved frequently from location to location, from job to job, and from relationship to relationship. After more than a year on the road with no fixed address, Brian returned to Illinois. He was unable to maintain stable employment or romantic relationships.

Within six months of returning to Illinois, Brian was arrested on suspicion of several unsolved burglaries. During questioning, he got in a shoving match with one of the detectives and was charged with assaulting a police officer. The charges were eventually dropped and Brian was released.

Brian became a well-known suspect to the police, who frequently arrested him on suspicion of committing burglaries and other crimes. In one altercation with the police, he received a broken nose and had to stay overnight in the hospital.

Brian's aberrant sexual behaviour continued to escalate. He was arrested several times on suspicion of sexual assault, but the victims were unable to identify Brian in line-ups.

He continued to drink heavily, as well as use a wide variety of illegal drugs. He supported himself through pawning proceeds from the burglaries he committed.

Just after his twenty-sixth birthday, Brian sexually assaulted a young woman and was arrested. The police considered him a suspect in other serious crimes. His public defender recommended he accept a plea deal for a ten- to fifteen-year sentence to avoid a trial and, potentially, a much longer prison term.

Brian was looking at spending the majority of his adult life behind bars.

Eric

Eric was sentenced to a supermax juvenile facility designed to manage the unmanageable. He was assigned his own cell. He enjoyed being alone. Eric spent just over a year in the facility before being released back into the community on his eighteenth birthday.

Eric relocated to California, where he found a job and enrolled in a local community college. Two years later, he switched to a four-year university and completed his undergraduate degree.

By age twenty-five, Eric had been employed for over three years. He had not committed a crime since being released from the juvenile supermax facility.

How was Eric's turnaround possible when he had exhibited such severe antisocial behaviour throughout his youth?

It turned out that Eric was a participant in an experimental treatment programme provided by the State of Wisconsin. Eric had received state-of-the-art intensive treatment at the supermax juvenile prison, known as the Mendota Juvenile Treatment Center, designed to stop him from continuing down the path of a lifetime of crime and incarceration.

The programme's success surprised even its creators.

The Mendota Juvenile Treatment Center

In the early 1990s, there was a national epidemic of juvenile violence in the United States. Rates of juvenile crime nearly doubled between 1980 and 1993, and there appeared to be no hope in sight to slow the flow of new juvenile offenders.[2]

Wisconsin, like most other states, was reeling from the massive flood of juvenile offenders. However, unlike other states, Wisconsin decided to conduct an experiment to determine if state-of-the-art treatment might stem the tide of the juvenile crime wave.

In 1995, the Wisconsin legislature established the Mendota Juvenile Treatment Center (MJTC) as part of a juvenile justice reform act. Located on the grounds of the Mendota Mental Health Center, the state's largest forensic psychiatric mental health hospital, MJTC was to be operated under the administrative code of the Department of Corrections but staffed by employees of the Department of Health Services.

The organizational structure was not created by accident. Because of its location on the grounds of the state psychiatric hospital, MJTC could evolve as a clinical-correctional hybrid in which physical security and mental health principles would operate in concert.

The Wisconsin legislature wanted to see results. Indeed, lawmakers required that a research psychologist be hired to conduct evaluations of MJTC and to follow treated youth upon their release back into the community to determine if the programme showed any benefit in terms of reduced recidivism compared to the regular standard of care in the juvenile justice system. Moreover, the research psychologist was to publish the results in the peer-reviewed literature – regardless of whether the programme worked.

MJTC started with forty-five beds. The clinical psychology staff included three psychologists and three social workers, for a ratio of two professional staff per fifteen youth. This was a very high ratio of staff to inmates. In one prison where I worked, there was a single psychologist responsible for the mental health care of over 550 men.

The programme also employed a full-time child psychiatrist. The day-to-day operation was managed by a psychiatric nurse. The nurse supervised all front-line staff, including those assigned with daily care, such as meals, hygiene, and education. Thus, the unit was administrated and operated by trained mental health professionals who were integrated into the security structure.

Because MJTC was located on the grounds of the psychiatric centre, the facility was able to avail itself of services from the hospital, including speech and language therapy, occupational therapy, dieticians, a chaplain, dentistry, and medical programmes.

The boys sent to MJTC were special, even by juvenile justice standards. Youth were selected for MJTC if the two other larger state juvenile correctional institutions were unable to manage them.

Nearly all the boys sent to MJTC had been deemed uncontrollable at the other institutions. They were the most severely violent youth in the state of Wisconsin. Indeed, the average youth sent to MJTC had over a dozen formally filed charges. Over 50 percent of the youth sent to MJTC had been convicted of a violent felony offense. The average Youth Psychopathy Checklist score was 28 out of 40, meaning the majority of youth at MJTC were in the severe range on the test. As reviewed in Chapter 4, the Youth Psychopathy Checklist scores range from 0 to 40, with the mean incarcerated youth scoring about 20. The Youth Psychopathy Checklist is the gold standard in assessing the callous and unemotional traits and impulsive behaviours that typify the highest-risk youth on a potential trajectory towards developing a full diagnosis of psychopathy as an adult.

In traditional correctional models, the state wields incredible power over the individual. Inmates who misbehave in a correctional environment are typically punished with more and more severe security measures, including isolation from the rest of the inmate population and prison staff up to twenty-four hours a day, seven days a week, three hundred sixty-five days a year.

These punitive measures are often seen by the inmate as something to be defied. The result is an escalating arms race in which the most violent or extreme side wins.

Violence by inmates begets only more violence. And as the famous Stanford Prison Experiment[3] showed, even prison guards are susceptible to adopting violent attitudes and behaviour towards inmates that they would never normally resort to in another environment. The Abu Ghraib prison scandal[4] in Iraq several years ago was a disturbing example of how guards can be adversely influenced by their environment and end up becoming violent themselves.

The Wisconsin legislature's decision to create MJTC was thrust upon staff at the Mendota Hospital with relatively little warning. In shaping the program, the founders examined all the peer-reviewed literature on psychopathy, the latest models of cognitive behavioural therapy for severely disruptive youth, and the publications on what had failed in the past to show any benefit to youth and adults in correctional settings.

The architects of MJTC were experts in treating violent psychiatric patients, patients who were psychotic (as opposed to psycho-pathic) and highly volatile, and who were further agitated by being locked in prison. The MJTC creators relied heavily upon their experiences working with this complicated population in designing the treatment model for MJTC.

It was a high-risk investment; the State of Wisconsin was betting that the psychologists at their flagship psychiatric hospital were up for the challenge. And a challenge it would be.

The leaders at Mendota responded by developing a radical new model that flipped the typical correctional model upside down. The treatment was designed to intervene in the self-defeating and mutually reinforcing interactions between the prison's need to control prisoner behaviour and the inmate's desire to resist.

The model was founded on the belief that prison deterrence and prisoners' defiant responses can become a vicious cycle. As the cycle repeats itself, prisoners give up more and more investment in convention, and their lives become 'compressed' as the use of punitive and restrictive sanctions increases. Eventually, individuals are so compressed that the only response left in their behavioural repertoire is violence.

The Decompression Model developed for MJTC was named after an ocean diver's slow ascent from the pressures of the deep. The programme sought to slowly uncompress the oppositional defiant behaviours and attitudes that pervaded the youth's psyche. The model created treatment aimed at developing basic prosocial bonds with the youth, to gradually 'decompress' their behaviours and attitudes, and to reorient their existing social skills towards minimal prosocial bonding.

Punishment was not effective for this segment of the youth prison population. It is sometimes difficult for the general public to understand that some people simply don't learn or benefit from punishment. Fear of punishment is something most of us take for granted. We learn to avoid punishment, which helps to form how we make moral decisions. However, as with all psychological phenomena, there is a bell-shaped distribution that ranges from those who are highly responsive to punishment to those who are highly resistant to learning from punishment. Most of the population falls somewhere in the middle of these two extremes. Unfortunately, when individuals with a low punishment temperament are thrust into the wrong circumstances, their resulting behaviour can be anti-social and violent. In extreme cases, such youth end up committing acts that result in their being placed in maximum-security juvenile prisons.

I came to appreciate the fact that punishment is not the only path to shaping behaviour early in my career. As I noted earlier, at university I volunteered to work on a project recording brain waves from captive killer whales. The trainers did not punish the whales for not doing as they were instructed. Rather, the trainers used positive re-inforcement practices, like feeding them salmon, to positively shape their behaviour in our experiment. Indeed, it seems silly to think of spanking a killer whale for not performing the way a trainer wishes, just as it is silly to harshly discipline a child who does not respond to punishment. It's just not an effective means of changing behaviour.

The first goal of the Decompression Model is to effectively ame-liorate the youth's violent institutional behaviour. As the youth's outlook improves, the next step is to conduct more psychiatric in-terventions, educational activities, and eventually offer more recre-ational activities.

Youth are monitored continuously throughout the day by the MJTC staff. Everyone contributes to monitoring the positive behav-iours exhibited by the youth, even if such behaviours are rare. By monitoring frequently, the staff can identify these positive behav-iours and reinforce them. The more the positive behaviours are rein-forced, the more likely they are to be repeated.

A critical component of the programme is that every staff mem-ber, from the cooks to the cleaning crew to the director of the pro-gramme, is trained and invested in the Decompression Model.

Reinforcement of the youth's prosocial behaviours is immediate and substantial. Using what is termed the 'Today/Tomorrow' pro-gramme, youth learn that if they are good today, they earn posi-tive reinforcement tomorrow. The rewards are graduated, in that if a youth is fantastically good today, the rewards will be fantastically good tomorrow. Rewards include things as small as candy bars to as much as the right to play video games in their cells. Psychologists call this type of positive reward structure *contingency management*.

Interestingly, brain scan studies have shown that both candy bars and video games are intrinsically rewarding – that is, the reward learning centres of the brain are engaged by both food and video games. The scientists at MJTC drew upon all the science available to them to improve their Decompression Model.

There was no way for the youth at MJTC to fail the programme. Youth were sent to MJTC because they could not be managed at other facilities. MJTC was their last resort. And the staff at MJTC refused to give up on a single youth in their care.

Youth who responded quickly to the Decompression Model were often transferred back to other juvenile institutions or, at the end of their sentences, released into the community. Youth who were more disruptive were more likely to spend the majority of their sentence at MJTC.

Does the MJTC Decompression Model Work?

The research psychologist at MJTC was required by the State of Wisconsin to determine if the Decompression Model was doing anything to alleviate the burden of crime in Wisconsin. In a series of studies, the research team followed over three hundred youth who had been admitted to the MJTC program over a five-year period. Next, youth treated at MJTC were compared to similar youth treated in the correctional system at other facilities.

The question was, Did the MJTC-treated youth commit fewer crimes following release into the community than the non-MJTC youth?

The results were nothing short of staggering.

Nearly all – 98 percent – of the non-MJTC youth were arrested for a new crime within four years of being released from juvenile prison. Only 64 percent of the MJTC youth were arrested during the same time window. In other words, the MJTC programme had resulted in a 34 percent reduction in recidivism.

But the really exciting results came from the reduction in violent crime. MJTC youth were more than 50 percent less likely to be convicted of a violent crime than non-MJTC youth after two years. At four years postrelease, MJTC youth were still 45 percent less likely than non-MJTC youth to have been convicted of a new violent felony.

MJTC did not just reduce violent crime by over 50 percent; it also reduced the severity of the type of violent crime committed. The non-MJTC youth killed sixteen people in the four-year period following release from prison. The MJTC youth did not commit a single homicide! That is a remarkable outcome.

The results of this research were submitted to scientific scrutiny and published in top peer-reviewed journals.[5]

I am more than a little excited about the results of the MJTC programme. It is the first time in history that a group of psychologists got together and specifically designed an intensive, one-on-one, cognitive behavioural therapy programme for youth with elevated psychopathic traits. And it worked.

Is Treatment Worth It?

The Wisconsin legislature had carved out the budget for the Mendota Juvenile Treatment Center (MJTC) from the state's allocation to the Department of Corrections. Such political manoeuvres can often leave those who have to endure the cut a little jaded. Those who receive the infusion of resources are in the onerous position of justifying the political process that led to the change in budgeting.

The research psychologist was charged with conducting an economic analysis to determine whether the MJTC program was cost-effective or not. He calculated the average cost per day of treatment at MJTC and other juvenile correctional settings in Wisconsin. Then he calculated the costs for criminal justice processing, including arrest, prosecution, and defense. Finally, he calculated the cost of imprisonment for those rearrested following release from prison by identifying the bed cost per day for each prison where those reconvicted were sentenced.

As expected, the state-of-the-art MJTC treatment was more than twice as expensive as the average daily bed cost in the regular juvenile prisons in Wisconsin. However, the research psychologist's findings showed that savings in the MJTC programme became evident within the first month that youth were sent to the facility.

The rate of institutional infractions dropped dramatically once youth were exposed to the Decompression Model. Because institutional infractions can lead to additional criminal charges and increases in the length of sentences, this saved money, as youth at MJTC were released from prison sooner than youth who accrued new charges and institutional infractions at the regular juvenile justice settings.

Additional significant savings occurred because the MJTC youth did not commit as many crimes upon release from prison as the comparison youth did. Recall that nearly all (98 percent) of the non-MJTC youth were back in prison within four years, whereas only 64 percent of MJTC youth committed new crimes.

Finally, the non-MJTC youth committed more severe violent crime, including those sixteen murders. Economists have estimated that murder is the most expensive crime to investigate, prosecute, and defend. And the individual convicted is likely to be sentenced to a maximum-security prison, which is an expensive type of prison.

Thus, in a detailed economic analysis, the research team at MJTC found that for every $10,000 the State of Wisconsin invested in MJTC treatment, it received over $70,000 in cost savings over a four-year period.[6] If the same $10,000 had been invested in the stock market, it would have yielded a return of just over $20,000.

Bottom line: The MJTC program was extremely cost-effective, largely because it reduced the future costs of incarceration.

Of course, the MJTC economic analysis examined costs only to the State of Wisconsin. It did not include additional costs that many economists believe are relevant to determining cost-benefit ratios. For example, crime has a significant financial impact on the victims, including lost wages and costs for replacing stolen property, victim services, and the medical costs associated with treatment for injuries sustained by victims of crime. We could add these costs to the savings gained by MJTC treatment.

The MJTC economic analysis also did not include the societal cost of the sixteen murders committed by the youth who didn't get treatment. Economists generally estimate the average American's economic contribution to society to be about $1,000,000 — so if a person is killed, it costs society a million dollars. The sum of the State of Wisconsin savings plus all the additional costs for victim services and losses due to the sixteen murders suggests the MJTC programme saved the State of Wisconsin tens of millions of dollars.

In addition, it is nearly impossible to place a numerical value on the emotional toll that crime has on society in general, particularly on the victims. How can one estimate the loss the sixteen families felt when their loved ones were murdered?

In other words, the MJTC programme has been an enormous benefit to the citizens of Wisconsin and a significant windfall for the State of Wisconsin budget. Indeed, I wish Wisconsin would allow me to invest my own money in MJTC. I'd sell my car and walk to work if Wisconsin would give me that kind of return on my investment.

With respect to Eric, he has been crime-free and gainfully employed for over five years since being treated and then released from MJTC. He writes the founders of MJTC that he has started to live a better life and he has not fallen back into the criminal lifestyle. He has stopped using drugs. He is seeking to further his education to help him get promoted at work. Eric credits his new outlook on life to the time he spent at MJTC.

Note that upon admission into MJTC, Eric had scored very high on the Youth Psychopathy Checklist. His attitudes and behaviour warranted the high score. However, given that he has ceased anti-social behaviour and is leading a normal life for a significant period of time, his adult Psychopathy Checklist score has to be adjusted. Indeed, based on all the new details provided by Eric, I would score him in the low range of the test now as an adult. I can't say for certain if the MJTC program *cured* Eric's personality, but it certainly demonstrated that Eric had the capacity for change. The amount of change we observed in Eric shows that the MJTC programme can alter the life of even the highest-risk youth.

Future of the Mendota Juvenile Treatment Center

As of 2013, the State of Wisconsin continues to invest in the Mendota Juvenile Treatment Center (MJTC) although the programme was recently cut from forty-five to thirty beds. The founders of MJTC continue to refine the Decompression Model to suit the needs of the highest-risk youth offenders. New studies continue to show the programme is working.

In 2012, my laboratory established a collaboration with the founders of MJTC. We deployed the mobile MRI to the grounds of the Mendota Hospital to study how the brains of youth participating in decompression therapy change over time. We want to know if decompression therapy changes the function and structure of the paralimbic system. The project is just getting started, and we are writing new proposals to funding agencies and potential donors who share our interest in understanding if decompression therapy offers a path to eventually cure psychopathy.

chapter 11

A Serial Killer Unmasked

fact: The FBI estimates there are as
many as fifty active serial killers in
the United States.

On average I receive about three e-mails a day from someone who is interested in learning more about psychopaths. I try to reply to all the e-mails and point people in the right direction to get answers to their questions. In the summer of 2009, I received an e-mail that was particularly intriguing. It turned out to be an e-mail that would change my life.

Dr Jim Cavanaugh, the director of the Isaac Ray Center in Chicago, Illinois, wanted to discuss a case with me that he was working on. I suggested a phone call, but he volunteered to fly to Albuquerque to meet me in person. I replied with some dates a couple of weeks out; he quickly wrote back that he wanted to see me the next day. I agreed, wondering what all the rush could be about.

We met for breakfast at my favourite coffee shop.

After some introductions, Jim got quickly to the point.

'I'm a forensic psychiatrist. I mostly work for the prosecution doing competency-to-stand-trial evaluations, psychiatric work-ups, and the like,' he started.

'But in the case I want to talk to you about, I've agreed to work on the defence side.'

'The client is named Mr Dugan. He is a serial killer. Perhaps you have heard of him?'

I searched my memory and then slowly shook my head as I exhausted my knowledge of serial killers. 'No, I don't know a serial killer named Dugan.'

'Then let me tell you a story.' Jim leaned back in his chair and looked around to make sure nobody was within earshot.

'In the early 1980s, Dugan sexually assaulted and killed at least three females – one adult and two girls. He was arrested in 1984 as a suspect in one of the murders and then he pleaded guilty to all three of them.

'Well' – Jim hesitated – 'the prosecutors let him plead guilty to two of the murders and he got two consecutive life sentences. But the last one . . . the last one is where the story gets a bit complicated.'

He took a drink from his coffee and looked at me intently. 'In 1983, a ten-year-old girl was taken from her home in the middle of the day, sexually assaulted, and murdered. The crime occurred in the town of Naperville, DuPage County, an affluent suburb just outside of Chicago. It was a horrific crime, and there was a massive manhunt to find the killer. The police were unable to find a suspect for several months. The public and media pressure on the police and the prosecutors to find the killer or killers was intense.

'Several months later, the police arrested three men for the murder. Two were found guilty and sentenced to death; the trial of the other ended in a hung jury.

'The verdict for the two convicted men was subsequently overturned by the Illinois Supreme Court because the prosecution had failed to turn over potentially exculpatory information.' Jim paused and took another sip of his coffee.

'The two men were retried, reconvicted, and resentenced to death again. Actually, I think one got a death sentence and the other life. The details don't really matter,' he said as he waved his hands.

It was clear he was trying to get through the story as fast as he could.

'And then, for the first time in the history of Illinois, the state Supreme Court overturned the death verdict, again.' He shook his head at the craziness of the case.

'But there was so much invested in convicting these two men that the prosecutors decided to charge them with capital murder for a third time. It was a big deal in the Chicago area.'

I tried to absorb the gravity of what he was telling me. This wasn't some Hollywood script – it was a true-crime story.

'The judge in the third trial said that the prosecution failed to prove its case and he dismissed the charges without the defence even having to put on its case. Many careers ended based on this case. It has been going on for over twenty years.

'Some have estimated that when you consider all the legal costs, it's the most expensive single case in US history.'

I interrupted. 'I'm confused, if they had Dugan's confession, why did they keep retrying the other guys?'

It made no sense in my mind. Crimes like this were nearly always committed by a sole perpetrator. I'd interviewed many men convicted of such crimes. Plus, Dugan had admitted to committing similar crimes, and he volunteered to confess to this crime. Certainly, he must have provided details about the crime that only the killer would know. The facts did not add up.

'I'll get to that. But it's at this point that the case gets interesting,' he added.

As if I wasn't already interested. He had me hooked at *serial killer*.

'Following the third trial, a special prosecutor was appointed to investigate the police officers and the prosecutors in the case. He conducted a year-long investigation, and as a result a grand jury indicted seven policemen and prosecutors on over forty counts of perjury, manipulating witnesses, and falsifying evidence. It was known as the DuPage 7 trial.'

'Holy shit,' I blurted out and then immediately apologized for my profanity. I have four cousins who are police officers in Phoenix, Arizona, and knew the stress that crimes like these can place on police departments. It can be very intense.

'The DuPage County board of directors voted to pay for the criminal defence of the policemen and prosecutors. But the people indicted were not allowed to plead guilty or plea-bargain. Apparently, it is against state law for county funds to be used for a criminal defence of its employees. The seven men went to trial, at a cost of millions of dollars to the county, and eventually they were all found not guilty. The jury went out and had beers with the defendants at a local bar following the verdict.'

'This story is just too impossible to believe.' I expected Aston Kutcher to jump out from behind the cashier and say I had just been *punked*.

Jim smiled. 'It is, isn't it?'

'The two men originally convicted of the crime spent ten years on death row as child rapist murders,' Jim stated.

We both knew what that meant. The two men would have been viewed by the other inmates as the lowest of the low in prison society. Other prisoners consider sex offenders, particularly those who committed sex crimes against children, to be beneath contempt. Those men, wrongly accused, must have been through hell on earth.

I asked, 'Did the two men sue DuPage County?'

'Yes,' he said. 'They were awarded a $3.4-million settlement from DuPage County.'

I ruminated on the figure. It wasn't enough. Three trials, two death sentences, ten years on death row – all because mistakes were made by the police and the prosecutors. My mind was swirling.

'Now, back to Mr Dugan. Even though he was not a suspect in the crime, he had tried to plead guilty to the ten-year-old girl's murder in 1984 when he pleaded guilty to the other two murders. But at this point the DuPage County prosecutors had already convicted the other two guys. So they claimed they did not believe Dugan's confession.'

Right, I thought. *The prosecutors were already invested in their convictions.*

'However, during the second retrial, the prosecution changed their theory of the crime and said that the two men committed the crime with Dugan. But Dugan was not charged. And Dugan refused to testify because the prosecution would not take the death penalty off the table. It's a very convoluted story.

'Finally, DNA analyses in 1995 proved that Dugan was responsible for the ten-year-old girl's murder.

'There is no statute of limitations on murder, of course, and a DuPage prosecutor recently charged Dugan with the killing of the ten-year-old girl.

'By all accounts Dugan is a psychopath. The prosecution is arguing that this makes him a danger to society and they are seeking the death penalty.

'Dugan has pleaded guilty to the crime and the only thing left to decide is the sentence. The decision of whether Dugan gets a third life sentence or a lethal injection will be up to a DuPage County jury. The prosecution has to demonstrate aggravating factors, such as the fact that he is a future danger to society, which is why they are raising the fact that Dugan is a psychopath. The defence has to put on a case for mitigation.'

And then Jim made his pitch. 'I'd like to know if you would be willing to come to Chicago to interview him and confirm that Dugan meets criteria for psychopathy and teach the defence team about the latest science on psychopaths. We may also want to get an MRI scan of Dugan if it is relevant to the defence mitigation strategy.'

I downed the rest of my quad espresso.

Jim continued, 'In 1982, I worked for the prosecution in the case of John Hinckley Jr, the man who tried to assassinate President Reagan. The defence used CAT scans of Hinckley's brain to support their diagnosis that he was psychotic and insane. The jury agreed with the defence and found him not guilty by reason of insanity. Hinckley was sent to a mental hospital.'

I was familiar with the Hinckley trial, but this was the first time I had met someone who had actually worked on the case. I reviewed the Hinckley case in my educational lectures to judges and lawyers because it was the first time in the United States that brain scans were used in court to support a psychiatric diagnosis.

Hinckley's delusion was that by killing President Reagan he would gain the attention of actress Jodie Foster (which he did) and eventually win her affection (which he did not).

The fluid-filled spaces of Hinckley's brain, known as the *sulci* and *ventricles*, were larger than normal. At the time, research showed that having enlarged sulci and ventricles was linked to schizophrenia. The defence argued Hinckley's enlarged sulci and ventricles were consistent with the diagnosis of schizophrenia.

But enlarged sulci and ventricles were not *diagnostic* of schizophrenia. That is, not everyone with schizophrenia had enlarged sulci and ventricles, and not everyone with enlarged sulci and ventricles had schizophrenia. The science at the time indicated that enlarged sulci and ventricles were *risk factors* for schizophrenia.

As Jim recounted his experience in the Hinckley trial, he said, 'The duelling psychiatrists for the defence and prosecution were arguing about whether Hinckley was psychotic or not. In my opinion, the brain scans tipped the scales in the favour of the defence.'

In the twenty-five years following the Reagan assassination attempt, it became clear Hinckley did indeed suffer from schizophrenia. His insanity verdict placed him in a mental hospital, where he got good mental health care. But at the time, using brain scan evidence to support a psychiatric diagnosis was controversial.

Today, using brain scans to bolster a diagnosis of schizophrenia is relatively simple. My laboratory team, in collaboration with my colleague Dr Vince Calhoun, have published many studies showing that we can differentiate patients with schizophrenia from healthy people, based on brain scans, with very high degrees of sensitivity and specificity. In 2012, we published a study showing we are 98 percent accurate at differentiating schizophrenia from healthy individuals based on brain scans.[1] We can even use brain images to differentiate different types of mental illnesses, like schizophrenia from psychotic bipolar disorder.[2]

Our work is designed to aid in early diagnosis of mental illness, so that we can help improve treatment for these patients. The legal system has recently become very interested in this work.

'We might want to do the MRI scans to illustrate to the jury what a psychopath's brain looks like to support the mitigation argument,' Jim stated.

Many scholars on the MacArthur Foundation Research Network on Law and Neuroscience felt that neuroscience was going to change everything about the legal system; others were not so sure. But one thing was clear: The project did not have any practising lawyers who were experts on the death penalty, and it certainly did not have any practising lawyers who were working on psychopathic serial killer cases. I figured that I might be able to learn what lawyers in the trenches needed and report back to the project members. I thought it might stimulate a lot of discussion.

'Yes,' I told Jim. 'I would be happy to meet with the legal team to teach them about psychopathy.'

I also told Jim, 'We might not need to get an MRI of Dugan's brain to help bolster the diagnosis of psychopathy, since the prosecution has already acknowledged that he met criteria for the disorder. But there might be other reasons to get an MRI, like a history of head injury. I would have to review the case file before making any recommendations.'

I'd spent the last several years reviewing hundreds of cases in which brain imaging evidence had been introduced in criminal proceedings. From bolstering psychiatric testimony to illustrating gross pathology like tumours or lesions, there were many reasons why a defence lawyer might want to have an MRI done of a client.

'Great,' Jim replied. 'I just have a few questions for you.

'This case is only about mitigation, not criminal responsibility. Dugan is responsible for the crime; he has already pleaded guilty. The only issue is sentencing. At sentencing the jury has to weigh the mitigating and aggravating evidence. Then the jury votes for life sentence or death sentence. A unanimous verdict must be reached for a death sentence.

'First, do you think that psychopathy constitutes an emotional disorder?' he asked.

'Yes. Psychopathy is associated with emotional deficits that are linked to impairment in nearly all domains of life; all my colleagues in the field would agree with that statement.' I hesitated. 'Well, all of them would agree that psychopaths have problems with emotion, but we may not all agree on the exact cause.'

'Is there peer-reviewed literature that supports these emotional deficits?'

'Yes, an extensive one,' I replied.

'Do you think that psychopaths have abnormal brains?'

'Yes,' I answered honestly. 'The science clearly indicates that psychopaths have *abnormal* brains.' As I answered, I wondered whether the legal system had a different definition of *abnormal* than the definition I used in my undergraduate courses.

'Do you think that psychopaths' brains are abnormal from birth?']

'Most of the evidence seems to point that way,' I said. I could already feel myself starting to hedge my answers. Jim's questions felt like a cross-examination. I wanted to make sure that I didn't overstate anything or say something that could be misinterpreted.

'In Illinois, if the defendant suffered from an extreme mental or emotional disorder at the time of the crime, it is considered a mitigating factor the jury must consider at sentencing,' Jim stated.

'Do you think that psychopathy is an emotional disorder?' he repeated.

'Yes. It is a lifelong emotional disorder,' I answered honestly again.

And then I told Jim about our work on emotional intelligence in psychopaths.

I explained, 'Emotional intelligence is the ability to identify, assess, and control the emotions of oneself, of others, and of groups. There are a variety of ways to assess emotional intelligence, including self-report tests, abilities-based tests, and interviews.

'While I was at Yale University, I worked with Dr Peter Salovey, who had spent his career developing the MSCEIT (pronounced 'mis-keet'), an abilities-based test for emotional intelligence. With Salovey's help, we implemented the MSCEIT in a large sample of prisoners, many of whom were psychopaths.[3] Psychopaths showed profound deficits in emotional intelligence compared to nonpsychopathic criminals. However, psychopaths had normal general intelligence (IQ) scores.'

Jim sat back and contemplated our findings. I could see that his mind had raced ahead to do a comparison of IQ with emotional intelligence.

'Jim, most of us know someone who has low IQ, and most people feel that if an individual with low IQ commits a crime, the legal system should consider them less responsible for their crime than someone who has normal IQ.

'As you probably know, in a landmark decision in 2002, *Atkins v. Virginia*, the United States Supreme Court ruled that individuals with low IQ were ineligible for the death penalty. The case before the Supreme Court was argued by University of New Mexico law professor Jim Ellis.

'When I moved to New Mexico, I met with Professor Ellis and showed him our work in emotional intelligence in psychopaths. Professor Ellis said that the same logic might apply to individuals with low emotional intelligence as it applies to individuals with low IQ with respect to death penalty decisions.'

As I finished my story, Jim raised his eyebrows and said, 'Wow. I'm glad that I came down to see you. Emotional intelligence research is a very important issue that we need to consider when evaluating the mitigation plan for Dugan's case, because if any case is going to be appealed all the way up to the United States Supreme Court and set a precedent, it's this one.'

'Jim, Isaac Ray believed that the emotional and intellectual sides of the psyche were equally important for the legal definition of *insanity*. Ray's thinking played a huge part in the trial of Charles Guiteau,' I added.

Jim's face blushed. 'Who was Charles Guiteau?' he asked. The forensic centre Jim directed was named after the prominent psychiatrist Isaac Ray, and he was embarrassed that he did not know all the cases Ray had influenced.

'Guiteau assassinated President Garfield in 1881. Ray's theories on *moral insanity*, the term used to define psychopathy at the time, were centre stage at his trial,' I said.

Jim contemplated the historical similarities between the Guiteau and Dugan cases. The historical tension between the legal system and the definition of *moral insanity* in the 1800s was being echoed by the new psychology and neuroscience of psychopathy in the 2000s.

It was just as many of the scholars working on the MacArthur Foundation Research Network on Law and Neuroscience had predicted.

Greenberg

Dr Jim Cavanaugh sent me binders of historical data about Dugan. There was so much detail in the files from Dugan's life that I felt like I didn't even need to interview him in order to complete the Psychopathy Checklist. The death penalty mitigation team had pretty much interviewed everyone he had ever come in contact with, including his family, childhood friends, neighbours, teachers, and employers.

But I was not going to let the opportunity to interview Dugan pass. I'd entered the field of abnormal psychology because I wanted to know how to stop serial killers like Ted Bundy before they ever got started, and now I was being given complete access to a serial killer not that dissimilar from Bundy. I wanted to learn everything about him; I wanted to know why he killed. I booked a flight to Chicago.

Lead counsel for the Dugan sentencing trial was defence lawyer Steve Greenberg. Greenberg wanted to meet for dinner. He said he would bring the latest mitigation research with him so I could review it before interviewing Dugan. We arranged to have dinner in downtown Chicago the night before my interview with Dugan.

One of my best friends, Ezra Friedman, an economist and lawyer, had just moved to Chicago and taken a job as a law professor at Northwestern University. I e-mailed Ezra and invited him to join us for steaks.

Ezra and I met with Greenberg at Gibson's Bar and Steakhouse on a busy Saturday night. The place was packed. As we approached the bar searching for Greenberg, I realized I had forgotten to make a dinner reservation.

Greenberg was easy to spot. He was the only one in the bar with both hands full. In his right hand he was carrying a large accordion-style binder with the latest mitigation details; in his left hand was a big martini.

After completing our introductions, Greenberg handed me the

binder and suggested we grab a table so we could talk in private about the case.

'What's the name on the reservation?' he asked.

'I forgot to make one,' I confessed.

Greenberg winked at me and walked up to the hostess desk.

As he approached the hostess, she smiled and gave Greenberg an embrace. She sat us at a corner table on the porch right along Rush Street. It was the best seat in the house.

The hostess returned a few seconds later with a fresh martini for Greenberg. 'On the house,' she said. Then she whispered something in his ear and he replied, 'Not a problem. Just let me know if he needs anything else.'

I looked over at Greenberg. He volunteered that he'd taken care of 'a little arrest for her cousin'.

As Greenberg explained the mitigation strategy, we were frequently interrupted by people walking along the street stopping to acknowledge Greenberg, thanking him for fixing this problem, thanking him for his advice about that problem. He always concluded each brief meeting with a hearty handshake.

Eventually, Greenberg stood up and moved chairs so that his back was towards the street, hoping to be less recognizable to the throngs of people walking by.

As our starters arrived, several professional basketball players were seated at an adjacent table. As if on cue, one of them leaned over to Greenberg and I heard him say, 'Thanks for taking care of that little problem for me.'

A bottle of wine arrived at the table. The waiter poured three glasses. Greenberg raised his glass to one of the NBA players in thanks.

Despite his prominence, Greenberg was a down-to-earth guy. I liked him immediately.

By the time our steaks were consumed, we were all on a first-name basis. And Steve had regaled us with too many stories to remember about the legal battles in the Dugan case. It was an epic saga that had brought the worst and best out of every aspect of the legal system.

Ezra knew several colleagues at Northwestern Law School who

had been involved in getting the two men who were wrongly convicted off death row.

After an evening of stimulating conversation and a large steak dinner, I said good night and retired to my hotel room to read the latest mitigation evidence. My first interview with Dugan was scheduled for Sunday morning.

As I reviewed the details of Dugan's crimes, I realized that it was going to be a tough interview. He had, at least three times, abducted females in broad daylight and taken them to a secluded location where he raped them and then murdered them by blunt force trauma. Two of the victims had been young girls; the other was a twenty-two-year-old nurse. They were the worst crimes imaginable.

The Dugan Interview

I awoke early, grabbed a quad espresso from the local coffee shop, and drove to the DuPage County Jail, where Dugan was being held. As I worked my way through security, I rehearsed several questions in my head that I wanted to ask Dugan during the psychopathy interview. I was placed in a comfortable interview room, and then the correctional officer left to retrieve Dugan.

The correctional officer returned with a middle-aged, grey-haired inmate dressed in a bright orange jumpsuit and white tennis shoes. He looked a little pale, but otherwise appeared to be in good shape. The officer removed the leg and arm cuffs, without asking me if I wanted the prisoner unshackled.

Once the officer left the room, the prisoner stood up and extended his hand across the table.

I stood and shook his extended hand and said to him, 'Mr Dugan, my name is Dr Kiehl, and your defence team asked me to interview you.'

He sat down and said, 'Please, call me Brian.'

Brian

I spent four hours with Brian Dugan during our first interview. We started by reviewing all the different facets of his life leading up to the time he was sentenced to life in prison for two of the murders he committed.

From his home life to his schooling, to his work history, to his romantic relationships, Brian's life before prison, as you have seen from previous chapters (yes, this is the Brian you first met in Chapter 6), was a mess, as corroborated by the reports I'd read from family, teachers, employers, and former girlfriends.

After covering all the basics, I had a pretty good idea that Brian was going to score very high on the Psychopathy Checklist. The traits were present in all aspects of his life. Brian was a textbook example of a psychopath.

But I wanted Brian to tell me why. Why had he committed those murders?

I had reviewed the different theories on why people commit sexual-based murders. By studying Brian's motivations and methods, I hoped to be able to help law enforcement catch other serial killers. I'd already identified a number of important clues from Brian's background information, and I was hoping Brian would reveal additional details that would provide insights into his methods and thinking.

One belief commonly held by psychologists is that adults who commit sexual-based crimes were once victims of sexual abuse as children. That is, there may be a cycle of offending when it comes to sexual-based crimes. While it's almost impossible to calculate the impact early trauma may have had on Brian or the thousands of other sexually abused children, it's hard to imagine that being sexually abused as a child would not impact some aspect of a person's constitution.

After covering the different areas of Brian's life, we had established a good rapport during the interview. So I started to ask him the tough questions about his sexual abuse as a child. But the questions ended up being harder for me to ask than for Brian to answer.

'Brian, the reports indicate that, as a youth, you were abducted by a man and forced to perform sex acts.'

'Yes. As a kid I got picked up by this guy driving a construction truck. He drove me out to a secluded area and made me give him a blow job,' Brian stated. 'Then he drove me back into town, gave me $20, dropped me off at a gas station, and took off.'

Brian almost chuckled when he told the story. The oddness of his behaviour derailed my train of thought. Deviating from my list of questions, I asked, 'Did that experience bother you?'

'No, not really. I mean, it wasn't a big deal. It didn't take very long and I got $20,' Brian said as he shrugged his shoulders.

The event didn't seem to have any impact on him emotionally, at least not in a way that he could articulate. He actually talked about it with some fondness, almost smiling. It was bizarre.

'And you identified the guy later from an article in the newspaper?' I asked.

'Yes. I told my boys' home supervisor about the incident when it happened, and then later I pointed out the guy from the picture in the newspaper when he was arrested for other crimes. Nothing ever came of it, though. The police never came, and no charges were ever filed.'

'You know who it was, though, don't you?' I asked.

'Yes. It was John Wayne Gacy. That's what I learned from reading the papers as the case developed. He killed lots of boys. So I guess I got lucky,' Brian said with a slight smile on his face.

John Wayne Gacy sexually assaulted and murdered at least thirty-three teenage boys in Illinois in the early 1970s. Brian was fourteen at the time of the reported incident. He described his assault by John Wayne Gacy as if he was telling me a story about putting his shoes on in the morning – with a complete lack of affect. His lack of emotion was as profound as I'd seen in any inmate I have ever interviewed.

Brian also admitted he had been sexually assaulted as a youth in the Menard Correctional Institution for boys.

'Did those experiences change your life in any way?'

Brian looked confused.

'Why would those experiences cause me problems?'

I inquired: 'Do you think those experiences had anything to do with the murders you committed?'

'Nope,' he replied.

He looked at me as if I was asking stupid questions. He had no conception that those early traumatic experiences might have influenced his adult behaviour.

There are several ways to look at Brian's flattened affect. On the one hand, Brian may have just been resilient to his childhood experiences. After all, the vast majority of youth who go through similar experiences don't end up becoming criminals, much less serial killers. Most people recover from, or are resilient to, early life stressors and do not develop severe mental problems as adults. It's how we survive. We cope and move on.

On the other hand, Brian might simply lack insight into how those events impacted his development. He definitely lacked insight into many domains of his life. But whether the lack of insight is a psychological mechanism for coping with the trauma, or a pre-existing neurobiological fault, remained to be determined.

The latest science would suggest the answer to this riddle is some combination of the two. According to genetics studies, at least 50 percent of Brian's constitution was established at birth. This high genetic loading for psychopathic traits, including lack of affect, may be the result of an underdeveloped paralimbic brain system, which put him on a steep trajectory towards becoming a psychopath as an adult. Moreover, the sexual abuse trauma he went through likely further dampened an already atrophied paralimbic circuitry, which made Brian even less able to experience emotion in a normal way.

The mitigation team had compiled a complete history of the emotional and behavioural problems Brian exhibited from a very early age. A flat, emotionless affect typified Brian's life from childhood through adolescence. And it was abundantly clear that the fifty-two-year-old man in front of me was completely lacking affect. Brian suffers from a chronic inability to appreciate the significance of his behaviour on others or their behaviour on him.

In fact, Brian's affect was so stilted that it took me a little by surprise. He was one of the rare inmates who appeared to be unable to appreciate or understand emotion on any level.

After I finished madly scribbling notes on my pad of paper, I moved on to my next line of questions, the planning for the murders. If his sexual abuse as a child was unrelated to why he committed the murders, then perhaps something else motivated his desire to kill.

'Did you plan out your crimes?' I asked.

'No, not really,' he stated. 'I never really planned any of them.'

'What about carrying duct tape, a tyre iron, and a knife?' I asked. Many serial killers kept a kill kit containing items they would need to abduct and murder the victim. Reports indicated the police had found those items in Brian's car when he was arrested.

'Oh, well, yeah. I had those things with me in case I needed them for burglaries, but I never really planned to kill someone; it just kind of happened,' he stated. 'And the duct tape was just for repairing the seats in my crappy car.'

Brian's words were consistent with the files. He appeared to be unable to plan any aspect of his life, including his criminal behaviour.

'So when you attacked women or tried to abduct them, you just picked them at random?'

'Yes. I used to drive around a lot looking for places to break in to do burglaries. I would smoke pot and then drive for hours. Sometimes I would not find a place that I liked and I would just go home. Other times I would hit five or six houses. But if I happened upon someone that I liked, then sometimes I just grabbed them and had sex with them.'

I'd prepared for this interview for weeks, but my mind kept racing ahead to new and different questions.

I found myself wondering which was more scary – worrying that someone is stalking you or knowing that there are people out there who will snatch you up completely at random.

Brian continued: 'Once I grabbed them, I would drive to a secluded place, have sex, and then take them home and let them go. One of the women called me to go out on a date again.'

I was dumbfounded. Did he really think that a woman would

want to have a relationship with him after being abducted and raped? Moreover, I just could not believe he would offer up his name and his phone number to someone he just raped.

'Brian, you gave the women you raped your phone number?' I asked incredulously.

'Yes. I would often give them my name and tell them to call me,' he said. Then he added, 'Seems kind of silly now, but at the time I thought we were having fun.'

Brian just didn't understand. I suspected he might set a new record for impaired emotional intelligence.

And then we got to the murders.

'Brian, why did you kill those girls?'

He looked up at me, shook his head, and said, 'I don't understand why. I wish I knew why I did a lot of things, but I don't.'

It was a completely unsatisfying answer.

But there it was. I had spent all this time planning to probe into the depths of his psyche, to try to find the events in his life that led him to rape and kill, to find out what motivated him. It had never occurred to me that he didn't know why.

I'd interviewed dozens of murderers, and many of them were serial killers. They all had reasons for killing. Their typical responses were *I was angry, I was after revenge, I was covering up a crime, I had to eliminate a witness, I killed because someone paid me,* or *I just wanted to see what it was like.* Some serial killers also murder because it turns them on. In fact, many serial killers are sadists in addition to being psychopaths. Sadism is a sexual-based disorder, or paraphilia. People who are sadists get sexually aroused by inflicting pain on others, and sometimes even causing death.

Brian was the first murderer I'd met who had no answer for why he killed. He didn't try to offer an excuse. He didn't try to blame anything or anyone else. He just didn't know why he did it.

'Brian, did you kill those girls because you wanted to eliminate them from being a witness against you in court?' I asked.

'No,' he said, shaking his head, 'I wasn't thinking about eliminating witnesses.'

He seemed to be telling the truth. After all, Brian had raped many more women than he killed. If he killed to eliminate witnesses, there would be a lot more dead victims. Instead, as perplexing as it was to me, Brian thought that the women he raped enjoyed it.

'Did you kill them for excitement?' I asked.

'No. I just did it quick as I could, I don't know why, it just kind of happened,' he repeated with a shrug of his shoulders.

His murders were motiveless. It made it harder to understand, and even more senseless. I took a break from the interview and excused myself to visit the toilet, so I could splash some cold water on my face.

While I was staring at my face in the prison toilet mirror, I was reminded of the seminal book on psychopaths that Dr Hervey Cleckley had written called *The Mask of Sanity*. I mentally rehearsed the symptoms Cleckley had identified that characterized psychopaths. One of the most perplexing symptoms Cleckley named was *Inadequately Motivated Antisocial Behaviour*.

It is a symptom that is almost an oxymoron. I had been trained to believe that all behaviour is motivated. But as I stared into that mirror, I truly understood for the first time what Cleckley meant when he included that symptom in describing psychopaths.

Psychopaths lack an ability to understand why they commit antisocial acts. They truly don't have the motivation that someone else might have.

I wasn't frustrated or angry anymore that Brian could not articulate why he committed those heinous crimes. I understood now that he lacked the capacity to understand. He was never going to be able to give a *normal* answer to the question of *why*.[4] There was never going to be a *logical* answer that he would argue to try to justify those terrible crimes.

When Brian was asked why he killed those girls, his mind just went blank. It's how the rest of us might feel if we were asked to solve Albert Einstein's field equations for the theory of general relativity – just plain emptiness.

Psychologists want to know what motivated Brian, and the millions of other psychopaths like him, to commit violent antisocial acts. But sometimes psychopaths do things without reason, without motivation. The rest of us search for some logic, albeit a morally twisted logic, that we can use to understand why. I came to accept the fact that there is no logical answer to many of the crimes people like Brian commit – as disappointing as that might seem.

As I stared into that mirror, I knew what I had to do next. Brian himself might not be able to tell me why he did the things he did, but his brain might.

I was overcome with an intense desire to use my mobile MRI to peer inside Brian's brain to find the answer to my burning question of *why*. I knew that Brian was unable to articulate his reasons for his actions, but unlocking the secrets of his brain might give me an answer that he could never articulate. Somewhere inside that head of his was a clue to how he became so disordered. And I was going to find it.

Bad Brains

The idea of unlocking the mysteries of Brian's crimes by peering into his brain may seem a bit funny. But many individuals afflicted with a mental disorder are unable to articulate why they acted the way they did.

One of my favourite examples is of an individual I'll call 'Brad'. Brad was a typical forty-year-old guy. He had no history of mental health problems, no history of problems at home, work, or school. He had a master's degree and worked as a schoolteacher, and he had recently married a woman who had a young daughter.

About two years ago, Brad developed a voracious interest in pornography. He spent hours collecting porn magazines and searching the Internet. He then started to visit massage parlours looking to solicit sex from the women working there. Brad later reported that his sex drive was so severe he felt he could hardly contain himself. Yet he felt this behaviour was immoral, so he concealed it from his wife. Then one evening he made sexual advances towards his prepubescent stepdaughter.

Brad's stepdaughter reported his behaviour to her mother, who had him arrested and removed from the home. Brad was diagnosed as a paedophile and prescribed a medicine (medroxyprogesterone acetate) designed to 'chemically castrate' him. Brad was a first-time offender, and a judge sentenced him to a secure inpatient sex addiction rehabilitation programme.

Brad's sexual behaviour continued to deteriorate while he was in the treatment programme. He propositioned other clients and even some of the staff. He was eventually expelled. Because Brad failed the rehabilitation programme, the judge was going to sentence him to prison. On the evening before he was to be transported to prison, he complained of a severe headache. The doctors ordered an MRI, and they found a huge tumour in the frontal lobes of Brad's brain, just above his eyes. This area, the orbital frontal cortex, is known to regulate impulse control and emotional reactions.

Brad's tumour was successfully removed. After recovering, he no longer reported any sexual interest in pornography, prostitution, or his stepdaughter. He was released from the hospital, and his wife took him back into their home. All went well for over a year. Then Brad developed inappropriate sexual thoughts again. He received another MRI and discovered the tumour had grown back. He was successfully treated again.[5]

My point is that it's easy to understand how gross brain abnormalities, like a tumour, can radically change a person's behaviour. Yet those afflicted are unable to articulate why they are behaving badly. During interviews prior to learning that he had a tumour, Brad just said that he had a strong sex drive. Brad did not know that the reason he wanted to have so much sex was that his brain was literally being crushed by a tumour. It was only after the tumour was discovered that Brad had an explanation for his inappropriate sexual behaviour.

Some legal scholars feel that Brad should still be found guilty of the crime of molesting his stepdaughter, regardless of whether his tumour caused him to do it. Indeed, some scholars argue Brad would be guilty if a little green alien in his head caused him to do it; that is, such scholars believe that if discovering the mechanism that causes a person to misbehave excuses that person from criminal responsibility, then there is no criminal responsibility, because all behaviour is caused by something. University of Pennsylvania law professor Stephen Morse refers to this causal conundrum as the 'fundamental psycholegal error'.

As interesting as the philosophical tenets of the doctrines of criminal responsibility are to debate, most people, including the US Supreme Court justices, argue that whereas individuals with brain differences or abnormalities might not warrant a complete exception from all criminal sanctions, the presence of such abnormalities does diminish their personal culpability. It was this language the US Supreme Court justices used when they eliminated the death penalty for youth (*Thompson v. Oklahoma* and *Roper v. Simmons*) and individuals with low IQ (*Atkins v. Virginia*).[6]

In essence, the Supreme Court's position is that populations of individuals with *different* behavioural profiles, and therefore *different* brain profiles, can be less culpable than healthy *normal* populations with respect to the death penalty. Juveniles who have not yet had time to reach full brain maturation and individuals with low IQ who have significant brain grey matter density are less culpable than normal adult populations because their brains are different. These positions are well supported by extensive behavioural evidence indicating youth and individuals with low IQ are fundamentally different from healthy adults, and the associated new neuroscience further bolsters this position.

The Supreme Court also has said that with respect to the doctrine of retribution – the interest in seeing that the offender gets his 'just desserts' or an eye-for-an-eye – the severity of the appropriate punishment necessarily depends on the culpability of the offender.[7] This seems intuitive. We punish youth less than adults, and we punish individuals with low IQs less than adults because we feel they are less responsible for their behaviour.

Further, bodies such as the American Psychiatric Association and the American Psychological Association have argued that developments in brain science continue to support that adolescents and individuals with low IQ should not receive severe sanctions because of their behavioural and brain differences relative to healthy adults. Proponents of this view continue to support legal cases that chip away at severe sentences, like the virtual death sentence

— life without the possibility of parole — for adolescent offenders. Supporters argue these severe sentences should also be abolished under the same logic as *Thompson, Roper,* and *Atkins.* Their efforts have been successful, and in a series of subsequent decisions the US Supreme Court has eliminated some of the most severe criminal sanctions for youth under certain circumstances. There are many additional cases pending where juvenile justice reformers are using new neuroscience to make arguments for reduced culpability for adolescents.

I find the logic in these arguments compelling and yet very challenging for society in general and the legal system in particular. The new neuroscience routinely demonstrates how mental illnesses are related to brain abnormalities. Neuroscience studies are routinely showing that posttraumatic stress disorder (PTSD), obsessive-compulsive disorder (OCD), schizophrenia, bipolar disorder, borderline personality disorder, and most other major mental illnesses are associated with impairments in brain structure, function, and connectivity. And one condition in particular that has piqued legal interest is the new neuroscience associated with psychopathy.

Brian's Brain

Lawyers who do death-sentence litigation are required to evaluate all aspects of a defendant's background to identify potential mitigating factors the jury will weigh in making their sentencing decision. If a lawyer fails to consider any potentially relevant mitigating factors, it can lead to successful appeals on the grounds that the counsel was ineffective in defending the client. Neither the mitigation lawyers nor the prosecution want that to happen because the former may end up disbarred and the latter may end up having to retry the case.

I gave my presentation to the Dugan legal team on the current and potential future uses of neuroscience in the courtroom. They quickly concluded there were many reasons to conduct an MRI exam on Brian.

Brian's life history was full of facts that might indicate brain abnormalities. Birth complications, early head trauma from banging his head against the wall to silence his persistent headaches, the severity and longevity of his headaches, numerous concussive events as a youth and at least one as an adult, and brain damage from alcohol and drug abuse were just some of the items that, in isolation or in combination, might be associated with neural trauma.

After reviewing my presentation on the latest neuroscience of psychopathy, the legal team wanted me to determine whether the structure and function of Brian's brain fit the profile of the hundreds of other psychopaths my laboratory had scanned. The legal team wanted to make the argument that psychopathy constituted a developmental disorder of emotion, and the new neuroscience might help make their point. This, the legal team felt, was something the jury needed to consider, and they would be remiss not to present this new science.

The legal team quickly got approval from the judge to get an MRI session for Brian. The judge, and even the prosecution, had agreed there were many reasons to have a look at his brain. I think the judge and the prosecution shared my interest in knowing if there was something different about Brian's brain.

After getting approval for the MRI scan, it fell to me to find a place where Brian could get the procedure completed. I told the legal team the MRI scanner we needed to conduct functional MRI studies would require special, and expensive, hardware that most average MRIs would not have available. Fortunately, my search was fairly easy to complete. I knew the Northwestern University medical campus in downtown Chicago had a research-dedicated Siemens MRI system like the one we used. So I called up the head physicist, Dr Todd Parrish.

Over the phone I explained to Todd that a legal team wanted to do a brain scan of a convicted serial killer.

'Cool,' he said. 'I've never looked at a serial killer's brain before.'

I asked Todd if he had any concerns about doing the case study. 'Nope,' he replied. 'It might liven things up around here.'

'We need to install some custom code on the MRI scanner and set up our tasks on your projection system,' I said.

'Not a problem,' Todd replied confidently.

Our custom code was pretty complex. It really dug deep into the Siemens software to pull out the raw imaging data at its earliest stage. I finished the call by detailing the type of custom imaging sequences and extraction code we would need to install on the Siemens MRI machine.

Todd is one of those rare physicists who can make an MRI machine hum any tune he desired. He installed all our custom sequences and code in a matter of minutes.

I was impressed.

Then Todd installed all the custom software needed to present the emotional and cognitive tasks Brian was going to perform while we collected images of how his brain functioned. He even set up our custom hardware for us so that we could monitor how Brian performed in real time.

I was even more impressed.

The primary purpose of the MRI scan was to make sure that Brian didn't have some sort of gross brain abnormality, like a cyst or tumour. The next question the MRI scan might answer was whether Brian had any visible brain abnormalities from the concussions he had suffered over the course of his life.

The legal team had also asked me to do a comparison of the grey matter density of Brian's brain to see if it fit the same pattern of abnormalities we have found in other psychopaths. And finally, I was going to examine whether the pattern of Brian's functional brain activity during emotion and attention tasks matched the psychopathic profile we had found in prior studies.

The question was: Did Brian show deficits in brain structure and function in the paralimbic regions of his brain?

After quite a few conference calls with the DuPage County sheriff's office, the legal team, and Northwestern security staff, we managed to arrange the transport of Brian Dugan from DuPage County Jail in Naperville thirty miles to downtown Chicago to be scanned at Northwestern's state-of-the-art Siemens MRI scanner.

Todd had asked that we arrange the scan session on a Saturday morning so the research area would be largely vacant. He didn't want graduate students walking into the MRI suite and bumping into a serial killer.

I flew to Chicago the night before Dugan's scan session and had dinner with Todd. We reviewed the protocols, and Todd assured me that everything was working fine.

Todd liked our code that interfaced with the Siemens software and pulled the MRI data off the scanner in 32-bit high-resolution format, and he asked if he could keep some of it. Siemens's engineers designed their MRI systems to collect extremely precise detail in the images. But in the United States, MRI scanners have adopted a low-resolution 12-bit radiological image standard called *DICOM*. The human eye can see only two hundred shades of grey, so radiologists thought they did not need all the extra resolution since they simply visually inspect the images to make their diagnosis. However, the algorithms we use to analyse functional brain imaging data can see many more shades of grey than the human eye, and the results are a lot better when we use the high-resolution data. I was planning to unleash the latest, most sophisticated algorithms in my neuroscience arsenal to find out what was going wrong in Brian Dugan's brain.

On Saturday, 5 September 2009, Brian Dugan arrived at the Northwestern University Center for Advanced MRI (CAMRI) for his brain scan, dressed in an orange jumpsuit and white tennis shoes. He was escorted by two DuPage County sheriff's deputies.

I was surprised there were only two deputies. Later I asked one of the deputies if it was standard procedure to send only two men on a transport of a serial killer. Brian's not a risk, they told me. He'd been in prison over twenty-five years and he didn't have a single institutional infraction. They considered him a model inmate.

Brian was very interested in the results from the brain scan. He had grown curious about what the scan might reveal about his brain. He asked a number of informed questions about the analyses and procedures we were about to complete. Apparently, he had been reading up on MRI and functional MRI from books and articles his legal team had given him, and he had become quite knowledgeable.

I had to remind myself that Brian's IQ was 122, in the very superior range, because every time I looked at him all I could think about was his low emotional IQ. As I had predicted, testing using the MSCEIT had revealed that Brian scored in the very low range on emotional intelligence.

I reviewed the scan session details with Brian, giving him practice trials for the tasks we were going to complete. After Brian completed the pilot testing, we took him into the MRI room and positioned him on the MRI table. With surprising efficiency Todd had Brian set up and ready to scan within just a few minutes. Todd adjusted the head coil and soft head restraint system to help Brian keep his head still during the MRI scan. We had only one item left to complete before starting to scan.

Todd handed Brian the emergency call bell and told him the device was to be used only in case he panicked and needed to get our attention in the control room. Brian calmly held the call bell in his left hand.

I was flooded with fond memories of Shock Richie and the antics he had caused on my first day scanning psychopaths more than a decade ago.

Brian just lay there calmly and gave a thumbs-up that he was ready to go. I looked at his eyes through the mirror attached to the top of the head coil. They were flat, showing no signs of anxiety.

I wondered whether Brian's age had anything to do with his calmness. Perhaps psychopaths age out of impulsivity? Maybe I will do a sabbatical year in Florida and see if there are any geriatric psychopaths around to study.

As we exited the MRI room the sheriff's deputy checked to see if it was secure and then followed us into the MRI control room.

'This is going to be interesting,' he stated.

It appeared everyone in the room, including the deputies and Brian, wanted to know what his brain looked like.

Todd sat down at the MRI console and set up the pulse sequences, checked on Brian through the intercom system, and the machine started up its familiar beeping and thumping.

The anticipation in the room was palpable.

As the first pictures of Brian's brain popped up on the Siemens console, Todd studied them and then proclaimed, 'Hmm. Looks pretty normal!'

The deputy next to him just chuckled and said, 'Not a chance that guy's got a normal brain.'

The rest of the MRI session went by quickly. Brian completed all the tasks that were assigned to him. However, I noticed he made a few curious mistakes during one of the emotional tasks.

I'd asked Brian to rate pictures based on the severity of the moral violation depicted in the scene. Examples of pictures with high immoral content included scenes depicting people rioting or a man shouting aggressively at a cowering child; pictures of scenes with low immoral content included bystanders looking at a car accident or a picture of a surgical procedure. The moral picture task had been developed by one of my postdoctoral fellows, Carla Harenski.

We had published a study showing that judgements of immoral pictures engage the amygdala in nonpsychopaths. But psychopaths fail to activate this paralimbic region of the brain. The normal boost the amygdala generates to help recognize moral content did not take place in psychopaths as it did in nonpsychopaths. It coincided with the psychopath's well-documented real-life impairments in moral decision making.[8]

Brian rated some of the stimuli with severe moral violations as far less immoral than everyone else rated them. In particular, Brian rated one picture depicting a man holding a knife against a woman's throat as low on moral severity. Later, during debriefing, Brian confessed that he knew most people would rate that picture as high in immoral content, but he said he was not sure it was *much* of a violation. As much as he tried, it was abundantly clear that Brian lacked the emotional connections that the rest of us take for granted.

We completed the scan session. Everything had gone perfectly; the code Todd had installed worked flawlessly.

As I was uploading the data to my server in New Mexico, Todd removed Brian from the MRI scanner. Brian was given a toilet break and fed a quick lunch. After the upload finished, I initiated my scripts to start processing the data. The algorithms would work all night to crunch the data. I had used code developed by my programmer to parallel process the data on multiple computers so the results would be ready by the time I got back to New Mexico the next day.

Once I was sure the data processing was under way, I backed up the information on a USB stick and grabbed my notes.

I had arranged to do another interview with Brian in a private office in the CAMRI building. As Brian sat back in the office chair, he looked down at the picture of his brain I'd given him.

I tried to decipher the expression on his face as Brian looked at the picture.

He looked up and noticed me staring at him. He asked, 'Is this the brain of a psychopath?'

I didn't need a picture of his brain to know I was sitting next to a psychopath. But I replied, 'You can't tell from a simple brain picture whether someone is a psychopath. You need a computer to do the detailed analyses of the MRI data in order to compare your brain scan against those of the other psychopaths. My laboratory has found that psychopaths have reduced density in the emotional areas of the brain.'

I continued, 'The brain is like a muscle, and some people's muscles are weaker than others. Weak muscles might be genetic, or they might develop because people don't exercise them. The brain is like that too.'

Brian nodded in agreement.

I went on. 'The latest science suggests psychopaths are born with weak tissue in the emotional control areas of the brain. Children as young as twelve years old who have symptoms like *lack of empathy, flat affect,* and *poor impulse control* have similar weak emotional brain areas, just as adult psychopaths do.'

The legal team had asked me to explain to Brian what the MRI scan might mean for his trial.

'Brian, the emotional brain tissue in psychopaths does not appear to be so weak or damaged as to make psychopaths have no power to control their behaviour. But it's likely to contribute to them making poor decisions. We just don't know exactly how weak brain tissue leads to emotional deficits, impulsivity, and the like.

'So this research doesn't mean psychopaths are necessarily less responsible for committing crimes. But in your case, the mitigation team wants to evaluate whether analyses of your brain support the argument you have an emotional disorder. They plan to argue that you've had this emotional disorder since you were a kid, perhaps even since birth. The latest science supports this argument. That's what the legal team wants to present to the jury.'

'This is really interesting for me,' Brian said. 'I've always known I was different. I wonder a lot why I did the things I did. Prison gives you a lot of time to think.'

He continued, 'I've read a lot about psychopathy since I interviewed with you last time. I read the list of symptoms, and I can see that I have almost all of them.'

This was the first time I'd ever had a psychopath recognize the symptoms in himself without any help.

'But I still don't feel any different. I mean, I really wish I felt different about what I did, but I just don't,' Brian said.

As he finished he looked at me with those flat eyes. I knew that he was trying, but he lacked the capacity. Brian didn't seem capable of appreciating the gravity of the crimes he committed. Even after decades of working with psychopaths, it was still extremely difficult for me to accept that fact.

I spent another hour or so probing Brian for details about his life, about how he thought about people, about how he thought about the crimes he had committed. He was in a very co-operative mood. He gave me a lot of insights into how psychopaths operate and how they think. I scribbled dozens of pages of notes.

I exited the interview room and went over to the MRI control room. Todd and the two sheriff's deputies were hunched over the computer looking at pictures of Brian's brain.

I heard Todd say, 'That looks a little weird . . . oh, wait. That's not abnormal. That's nothing. Sorry.'

And with that the three of them leaned back in unison and took another look at the screen from a distance.

'FIND ANYTHING?' I said with a loud voice, startling all three of them.

'Jesus!' Todd exclaimed. 'How can you sneak up on someone when there is a serial killer in the building?'

The deputies and I had a nice laugh about that one.

As the deputies fetched Brian, I asked Todd if he had found anything abnormal in Brian's brain.

'Nope. Looks pretty good, actually. I can't wait to see what the VBM analysis yields.'

VBM stood for *voxel-based morphometry*, the density analyses my computers back in New Mexico were cranking out as we spoke.

'Me too,' I answered.

Results

The next morning while I was waiting at the airport for my return flight to depart, I logged in to my computer servers to check on the status of the analysis of Brian Dugan's MRI datasets. The brain density analysis was nearing completion; the four analyses of the functional MRI data were only about 20 percent done. I sipped my espresso and watched the status bar creep towards completion on the density analyses of Brian's paralimbic system.

Unfortunately, my flight was boarding. I'd have to wait until I was back in New Mexico to get the results. But before I closed up my computer, I e-mailed one of my graduate students, Lora Cope, and asked her to analyse a 'case study' for me. I told her where the case study MRI data resided on the servers and the exact analyses I wanted her to do. I wanted an independent analysis to verify my results.

During the flight home, I tried to decipher the notes I had hastily scribbled during my interview with Brian. I planned to type them up while they were still fresh in my mind. I balanced my computer on my lap, and my notes were sticking out from the seat pocket on the seat in front of me.

I was making good progress typing up my notes until the young woman in the middle seat got up and bumped my precariously placed notes, spilling them onto the floor. She quickly apologized and headed to the back of the plane towards the lavatory.

It took a bit of time to reorganize them and get back to work. I was just about to settle back into typing again when the flight attendant announced we were landing shortly. I closed my laptop and packed up my bag.

As the wheels touched down, I clicked on my smartphone and logged in to my computer servers. The density analysis was complete. I could not wait to get home to review the results. I raced out of the plane straight towards the security exit and the parking garage to get my car.

As I passed security, I felt a strong hand grab my left arm. I turned and came face-to-face with a uniformed police officer. The look on his face was sheer terror. Confused, I turned and looked around for what might be causing his alarm.

'Would you please come with me, sir?' he asked.

'What's the problem, officer?' I was perplexed.

'Would you please just come with me, sir?' he repeated.

'Umm, sure,' I managed to choke out. I really wanted to get home to look at the results of Brian's brain scan. But I realized I better not mess with the airport police.

We proceeded to a small security office where the officer motioned for me to sit in the chair. I noticed he sat in the chair closest to the exit. Another officer, much larger than the first one, entered the room.

'May I search your bag, sir?' the large one asked.

'Of course,' I replied.

The large officer proceeded to open my computer bag and remove its contents. I am a bit of a packrat. I had all sorts of receipts from various academic trips that I had not turned in yet for reimbursement, as well as journal articles that I was in the middle of reading, and pages and pages of notes from the Dugan interview. As the officer pulled the papers out of my bag, a few of them fell to the floor. I instinctively quickly reached to catch them, and both officers jumped back in unison.

The big one put his hand out as if to block an attack while the smaller one took a quick step towards the exit door. It was an awkward moment.

As I retreated to my chair I asked: 'What's going on, guys?'

'Please, sir, just stay seated while we complete the search,' was all the large officer said.

I sat back and tried to relax.

The large officer handed the stack of papers from my bag to the smaller officer to review, and then he removed a prescription pill bottle from my travel bag. It was a muscle relaxant in case my back acted up. He looked at it curiously and then replaced it in the bag and continued his search.

The small officer stood up and exited the room, taking the stack of my papers with him. I didn't object; I figured he was going to test them for explosive residue or something.

The large officer completed his search and then placed my bags on the floor for me to reclaim. He excused himself and I was left alone in the security office.

A few minutes elapsed.

A new officer came in and sat down; he had my stack of papers with him. He was followed by the two other officers, who took up positions along the wall farthest from me. Their arms were crossed in a pose of strength.

'Can you tell me your name, please, sir?' asked the new officer.

'Dr Kent Kiehl,' I answered.

'I'd like to ask you about the content of these notes,' he said. 'A passenger on your flight notified the authorities that you appeared to be typing out a confession to a series of murders. We want to know if you want to confess to us.'

I had written my notes for the Dugan interview in first person, just as Brian had relayed the details to me. Brian had given me detailed information about the murders and his other crimes.

The woman in the middle seat must have been reading what I was typing on the computer. I realized just then that she had not returned to her seat before we landed.

The cops looked like they had just made the arrest of the century.

My detailed notes read just like a confession, which was a lot more evidence than the prosecutors had on the two men who had been wrongly convicted.

I motioned slowly to the officers that I was going to reach into my pocket. I did not want to startle them again. I slowly withdrew my wallet and removed Steve Greenberg's business card. I slid it across the table and said: 'Gentlemen, I'm afraid that I am unable to discuss with you the contents of the notes. I'd like to call my attorney.' I tapped on the business card.

The new officer was clearly frustrated. He turned and looked back to the other two officers and motioned for them to leave the room. He stood up.

'You can have five minutes for a call, but you are not leaving without telling us about those notes,' he said emphatically as he left the room in a huff of frustration.

I needed to ask Greenberg whether the details of Brian's murders were confidential or whether I could tell the officers the truth.

Steve picked up the phone quickly. I was relieved to hear his voice.

'What's up, Doc?' he said in a heckling voice.

'You are not going to believe this,' I started. I told him the situation, that a woman sitting next to me on the plane thought I was writing out a confession to murders I had committed and had told the police at the airport. 'I'm in an interrogation room right now. Is it okay to tell them my notes are about Dugan or do I have to keep them confidential?'

Steve's laughter on the other end of the phone was predictable, but not very funny to me at the moment.

'Normally, I would tell a client in your position to keep your mouth shut,' Greenberg said, still laughing, 'but in your case, just tell the cops where the notes came from. Dugan has already confessed to the three murders and pleaded guilty. There is nothing to hide. Besides, we gave you permission to publish Dugan's case. Go ahead and tell the cops.' And then he finished by saying, 'And if they don't believe you, then give me another call and I'll fly down and try to get you out of jail.' The last statement trailed off as he broke out laughing again. And then he hung up.

I stood up and went over to the door and knocked, and the officer who had been questioning me walked in.

'I've been given permission to tell you about the notes,' I said flatly. 'I study psychopaths for a living, and I just returned from Chicago where I was interviewing a psychopathic serial killer named Brian Dugan. Those notes are his words from an interview I conducted yesterday.' I pulled the Dugan file from my bag and showed the officer the court order allowing the transport and interview of Dugan.

He looked at the court order suspiciously.

I continued, 'I have some media clippings on my computer that I can show you that describe the crimes Dugan has committed.' I reached into my bag and pulled out my laptop. The officer took a seat across from me at the table.

I pulled up the latest articles from the *Chicago Tribune* that detailed the Dugan case. I spun the laptop around and showed the officer. He quickly skimmed the headlines and then excused himself and left the room.

I packed up my laptop and papers, and I placed my computer bag on top of my carry-on bag. I wanted to get out of the airport as quickly as possible.

The officer returned and said the most comforting words I had heard in a long time: 'Dr Kiehl, you are free to go.'

And then he continued, 'Please don't type out your notes on such cases on airplanes again.'

I promised him that I would not.

I drove home staying under the speed limit the entire way.

Results Part II

Monday morning I settled into my home office to begin the review of Brian's brain density analyses. I entered the secure network and proceeded to pull up the results. I planned to complete three checks this morning. The first step was to conduct a quality control assessment to make sure all the processing had been completed without error. The next step was to conduct an *internal physiological control*[9] analysis, comparing Brian's data with control subjects' in brain regions that I did not expect to show any impairment; that is, Brian should show normal brain tissue in regions outside the paralimbic system. Finally, the last step was to examine whether the paralimbic regions of Brian's brain were atrophied, as they were in our studies of other psychopaths.[10]

The first step took only a few minutes to complete. All the stages of image analyses had proceeded without any problems.

For the second step, I programmed the computer to randomly select a dozen brain regions outside the paralimbic system. I extracted density values from those regions of Brian's brain and compared them to my database of over one thousand control subjects.[11] Brian's density values for the control regions were well within the normal range.

In the final step, I extracted the density values for the paralimbic regions of Brian's brain and compared them to the normal database and the database of other psychopaths we had scanned.

The density measures from Brian's paralimbic system were strikingly consistent with our findings from other psychopaths. In every region of the paralimbic system, Brian showed the same pattern of atrophy we had seen in other severe psychopaths. Brian had scored in the 99th percentile on the Psychopathy Checklist, and his brain data fit within that percentile on grey matter density. Brian's paralimbic grey matter values were even more atrophied than most psychopaths'.

I downed the rest of my coffee and sat back in my chair as I processed the results. I'd looked at hundreds of psychopaths' brains in my career, but the consistency of their brain abnormalities never ceased to amaze me.

Redundancy

As I passed the offices of my students and postdocs at the Mind Institute, I noticed light emanating from Lora Cope's office. I slowed down and gave the door a soft kick with my foot, announcing my presence. Startled, Lora jumped up, and then looked at me sheepishly.

'You got me again.'

I had a notorious habit of startling my lab staff. It was a vestige of my years growing up with three sisters. My parents had raised me to never be physically aggressive with my siblings, so many of our childhood altercations took the form of psychological digs and jabs.

My sisters had devised numerous ways to get me. My retaliation had come in devising devious ways of startling my sisters when they least expected it. My behaviour carried through into my adult life.

I always felt a bit of relief when my staff startled if I gave them a quick scare. After all, psychopaths have a very small startle response. It's kind of my little psychopath test to conduct on people in my lab.

After she recovered, Lora looked at me and said, 'The paralimbic system of the brain scan you sent me is the most atrophied that I have ever seen.'

Lora had analysed the brains of hundreds of psychopaths over the past several years, and she was one of the world's leading experts in the nuances of brain density analyses of psychopaths.

'I've worked up some figures for you on the data; I'll send them to you shortly. And the functional imaging results will be ready by the end of the day,' she said.

The results from functional MRI tasks collected on Brian confirmed what I had expected. Like other psychopaths, Brian failed to show the normal engagement of paralimbic structures during processing of emotional stimuli. In addition, in the attention Oddball Task, Brian failed to show the engagement of the orienting response system, which included the majority of the paralimbic system, results that were consistent with the peer-reviewed literature on psychopaths.

The analyses of Brian's brain had revealed how he was different from the rest of us. Indeed, the pattern of results across the various brain analyses was amazingly consistent with the latest science of psychopathy.

I wrote up the report and submitted it to the legal team, outlining how Brian's brain imaging data fit within the known abnormalities in psychopaths. Next, I created a presentation to educate the jury on the latest neuroscience of psychopathy and how Brian's Psychopathy Checklist score and brain data fit within that literature.

Sentencing Trial

During Brian's sentencing trial, the judge granted permission for me to observe the other members of the team as they presented Brian's history. Brian's entire mental health records were put up for the jury to review. A psychologist testified about Brian's behavioural and affective problems as a youth. Dr Jim Cavanaugh offered his clinical opinion regarding Brian's emotional problems. Many days of testimony were offered to the jury. The final piece would be my presentation.

Jury Presentation

The morning of my presentation to the jury, I got up early and walked over to the coffee shop to order an espresso. After dressing and looking over my PowerPoint presentation one more time, I headed down to the lobby to meet Greenberg for the short ride to the courthouse.

Inside the courtroom, I connected my computer to the projection system that would display my presentation for the jury. It was an enormous screen, larger than most found in academic classrooms. I went through my slides again, making sure they were in the right order.

The judge called in the jury.

I was called to the witness stand and I gave my presentation. The jurists listened intently as I told them about the latest science of psychopaths. One young man in the front row of the jury box was jotting down a lot of notes.

I explained that Brian's brain imaging data were consistent with what was known about deficits in psychopathy. Several members of the jury nodded in unison.

Everyone in the room seemed to know something was wrong with psychopaths, and with Brian.

At the end of the presentation, I was asked by the mitigation lawyer why I chose to make a career studying psychopaths.

I truthfully replied, 'So we can develop better treatments for psychopaths and prevent them from ever committing crimes.'

After the jury was excused, I got down from the witness box, and the prosecutor came up to me and shook my hand, saying I was an excellent expert, and I had a long career ahead of me if I wanted to do this for a living.

I told him that if I had my way, I'd never testify in another case like this. In my ideal world, Brian would have received treatment as a youth and perhaps, just perhaps, we could have prevented his crimes from ever taking place. The prosecutor nodded in agreement.

Verdict

The testimony complete, the mitigation team and prosecution gave their closing arguments. The judge ordered the jury sequestered to deliberate on their sentencing verdict. The court officer drove the jury in a van to an undisclosed hotel.

The mitigation team met for a final dinner, to congratulate one another on the hard work they had put in over the past year. I'm certain the prosecution team was doing the same.

One of the lawyers said that DuPage County had never had a jury go out more than two hours in a death penalty case. In fact, he said, a DuPage County jury had *never* returned a life sentence in a sexual-based homicide.

I returned to New Mexico the following morning as the jury started its second day of deliberations. Then, on the third day, I got a single-word text from Greenberg: 'Death.'

It was not an unexpected verdict.

Outcome

It was 11 p.m., about six hours after the jury had returned a sentence of death for Brian Dugan, when my phone rang. The caller ID indicated it was Greenberg.

'Hi, Steve,' I said as I answered the call.

'You are not going to believe this!' Steve exclaimed. 'I have to write up a subpoena to serve on the judge in the Dugan trial tomorrow,' he stated.

'What are you talking about?' I asked.

'Haven't you been watching the news?' he asked.

'No. Sorry. What's going on?' I asked again.

'A jury member from the Dugan trial told a reporter that a signed verdict of a life sentence had been turned in to the bailiff and handed to the judge. But the judge never told counsel about that signed verdict.'

'Oh, my god,' I stated.

'Exactly! Tomorrow I have to go in and subpoena the judge to preserve that verdict. I have no idea if this has ever happened before in the history of law,' Greenberg said. I could feel him shaking his head on the other side of the phone.

'This is just unbelievable,' he continued. 'This whole case will be overturned, and we will have to start over. It will cost the state millions.'

Subsequent interviews with the jury indicated that the initial vote had been four of twelve jurors recommending a life sentence for Dugan. After many hours of deliberation, the jury had signed a verdict sheet with two abstaining, two voting for life, and eight voting for death. The jury had handed the signed verdict sheet to the court bailiff, who then gave it to the judge.

The DuPage County Courthouse had been transformed into a maximum-security setting in preparation for the final verdict in this twenty-five-year-old case. The lawyers were escorted under armed guard through a secure entrance to the building. Bulletproof glass shields were installed in the courtroom to protect the jury and the legal teams in case someone smuggled a firearm through the metal detectors. SWAT teams were deployed for crowd control. Hundreds of protesters gathered outside the courthouse.

However, before the verdict could be read, one or more of the jurors requested permission to restart deliberations. The judge notified the defence and prosecution that the jury was resuming deliberations, and then he sequestered the jurors for the night in a hotel. The judge did not give judicial notice that a signed verdict had been received. It is unclear whether the judge read the verdict on the sheet or not, and the bailiff claimed he had not been able to read it because he did not have his glasses.

The following morning the jury deliberated for two hours and then returned a second verdict: death.

Until the television interview that evening, Greenberg had had no idea that more than one signed verdict was in existence. The next morning Greenberg appeared before the judge and secured the original verdict sheet voting for a life sentence.

An immediate appeal was filed with the Illinois Supreme Court. The mitigation team felt that the case was going to be overturned and a new sentencing trial would be ordered.

It might be the most convoluted murder case in US history.

Final Chapter

As Dugan's appeal worked its way through the judicial hierarchy, another legislative milestone occurred in Illinois. Citing the large number of wrongly convicted men on death row and the enormous cost of litigating death penalty cases, Governor Pat Quinn signed into law a bill to abolish the death penalty in Illinois on 9 March 2011. And since the bill was not retroactive, Quinn commuted the sentences of the fifteen men on death row, including Brian Dugan, to life without the possibility of parole.

Dugan could have kept the courts tied up for years litigating the errors in his case, but in what is likely one of the only altruistic acts of his life, he voluntarily dropped all his appeals. He was removed from the death row prison and sent back to his home institution, where he will spend the remainder of his life.

Epilogue

The Brian Dugan case illustrates how a heinous crime can alter the lives of so many people. The abduction, rape, and murder of ten-year-old Jeanine Nicarico forever changed the town of Naperville in

DuPage County, Illinois. The crime brought out the best, and worst, from police officers, prosecutors, media, politicians, public defenders, defence lawyers, judges, law professors, and the general public.

The case also illustrates the ways the new neuroscience of individuals with severe emotional and behavioural disturbances challenges our beliefs about free will, punishment, and justice.

Since my experience with Brian, people often ask me if I am pro death penalty or anti death penalty. I answer that I am 'pro prevention'. Treatment programmes, informed by the best evidence-based research, are one way to curb the tide of crime and violence that plagues society. For if society can implement cost-effective treatment programs like that at the pioneering Mendota Juvenile Treatment Center (MJTC), perhaps we can prevent crimes that lead to victims like Jeanine. Indeed, by all accounts MJTC has been a resounding success. I have the utmost respect for the founders, Drs Michael Caldwell and Greg Van Rybroek. They are true pioneers. They ignored centuries of dogma that psychopaths could not be treated or managed. They developed a treatment programme that has saved millions of dollars and untold numbers of lives. They are the true psychopath whisperers.

I'd like to close by thanking the reader for allowing me to share my journey into the mind and brain of the psychopath. I hope this book will stimulate discussion about how to make a better and safer society for everyone. As for me, well, I have more work to do — it's time for me to go back to prison.

Acknowledgements

To my wife and daughter, friends and family, current and former staff and students, collaborators and colleagues, teachers and mentors, and my agent and editor – this book would not have been possible without you. Drinks are on me.

Notes

Chapter 1: Maximum Security

1. Dr Robert Hare first published the Psychopathy Checklist in 1980: Hare, R. D. (1980). A research scale for the assessment of psychopathy in criminal populations. *Personality & Individual Differences* 1 (2), 111–119. He subsequently published the manual for the Hare Psychopathy Checklist in 1991 and a revision of the manuscript in 2003 with updated research findings: Hare, R. D. (1991). *Manual for the Hare Psychopathy Checklist-Revised*. Toronto: Multi-Health Systems; Hare, R. D. (2003). *Manual for the Hare Psychopathy Checklist-Revised* (2nd ed.). Toronto: Multi-Health Systems.

2. Porter, S., Brinke, L., & Wilson, K. (2009). Crime profiles and conditional release performance of psychopathic and non-psychopathic sexual offenders. *Legal and Criminological Psychology* 14 (1), 109–118.

3. Monahan, J. D. (1981). *The Clinical Prediction of Violent Behavior.* Washington, DC: Government Printing Office, pp. 47–49.

4. Hart, S. D., Kropp, P. R., & Hare, R. D. (1988). Performance of male psychopaths following conditional release from prison. *Journal of Consulting & Clinical Psychology* 56 (2), 227–232; Hare, R. D., & McPherson, L. M. (1984). Violent and aggressive behavior by criminal psychopaths. Special Issue: Empirical approaches to law and psychiatry. *International Journal of Law & Psychiatry* 7 (1), 35–50.

Chapter 2: Suffering Souls

1. As of 1 January 2013, the population of the world is approximately 7,100,000,000 (3,550,000,000 males). The rate of psychopathy is between one half of a percent and 1 percent in males or 17,750,000 to 35,500,000 (average 26,625,000). In women, the rate of psychopathy is about one

tenth of 1 percent or 1,775,000 to 3,550,000 (average 2,662,500). Taking the average number of male and female psychopaths from the estimates above, we get a total of 29,287,500 psychopaths in the world today. http://www.usarightnow.com

2. Seabrook, J. (2008). Suffering souls. *The New Yorker*, November 10, 2008.

3. Koch, J. L. A. (1888). *Kurzgefasster Leitfaden der Psychiatrie mit besonderer Rucksichtnahme auf die Bedurfnisse der Studierenden, der praktischen Ärzte und der Gerichtsärzte.* Ravensburg: Dorn.

4. http://en.wikipedia.org/wiki/Book_of_Deuteronomy; written in Jerusalem in the seventh century BCE in the context of religious reforms advanced by King Josiah (reigned 641–609 BCE) (last visited 11/16/11).

5. Widiger, T. A., Corbitt, E. M., & Million, T. (1991). Antisocial personality disorders. In A. Tasman & M. Riba (Eds.), *Review of Psychiatry* (vol. 11). Washington, DC: American Psychiatric Press.

6. Murphy, J. M. (1976). Psychiatric labeling in cross-cultural perspective. *Science* 191 (4231), 1019–1028.

7. Widiger et al., Antisocial personality disorders.

8. Pinel, P. (1801). *Abhandlung über Geisteverirrunger oder Manie.* Wien, Austria: Carl Schaumburg.

9. Very rarely will a patient with a psychotic disorder, like schizophrenia, also meet criteria for psychopathy. I've interviewed over one hundred patients with schizophrenia who have been convicted of violent crimes, and only one met criteria for psychopathy using the Psychopathy Checklist. This patient was very odd, and the delusions that pervaded his thinking inflated his ego, made him grandiose, and contributed to the crimes he committed. He did not present like the typical psychopath during our interview, but nevertheless sticking to the rigorous procedure for assessing psychopathy, this patient did score over the diagnostic threshold. It's been my experience (also shared by many other scientists) that psychopaths rarely experience psychotic symptoms. The one caveat is that psychopaths sometimes experiment with psychedelic drugs. But we don't consider short-term psychotic symptoms associated with drug use to be of the same quality and severity as such symptoms experienced in patients with schizophrenia.

10. Rush, B. (1827). *Medical Inquiries and Observations Upon the Diseases of the Mind.* Philadelphia: Jonkoping.

11. Prichard, J. C. (1837). *A Treatise on Insanity and Other Disorders Affecting the Mind.* London: Haswell, Barrington, and Haswell.

12. Maudsley, H., & Harry Houdini Collection (Library of Congress). (1875). *Responsibility in Mental Disease.* New York: D. Appleton and Co.

13. Ray's *A Treatise on the Medical Jurisprudence of Insanity* (1838) was the first systematic treatment in English of the problem of mental illness in relation to crime and punishment (cf. Overholser, 1962). Ray, I. (1838). *A Treatise on the Medical Jurisprudence of Insanity*. Boston: Freeman and Bolles.

14. It's noteworthy that the term *psychopathy* may have first been used by the Austrian physician and author Ernst von Feuchtersleben (1806–1849), who in 1845 wrote the first psychiatric textbook in Austria, *The Principles of Medical Psychology* (English translation 1846). However, it wasn't until Koch's work that the term became popular.

15. Birnbaum, C. (1909). Über Psychopathische Personlichkeiten. Eine psychopathologische Studie. *Grenzfragen des Nerven- und Seelenlebrens*. Wiesbaden.

16. The full list of symptoms articulated at Karpman's first meeting on psychopathy were:

    1. Despite normal intellectual functions, behaviour is radically abnormal.
    2. Mendacity – inability to keep to the truth, without any motive for lying. Also referred to as pathological lying or untruthfulness.
    3. Lack insight into how their behaviour impacts others.
    4. Behaviour resistant to change.
    5. Punishment ineffective in changing behaviour.
    6. No evidence of psychotic symptoms (no hallucinations, delusions, or aberrant thinking); this differentiates the psychopath from the individual with schizophrenia or bipolar disorder.
    7. Profound failure in emotional domains; described variously as emotional instability, emotional flatness, reduced expression and range of emotions, emotionally disconnected from others; inability to form long-lasting emotional bonds with people.
    8. Inability to feel empathy or love.
    9. Guiltlessness – feelings of guilt seem to be lost or absent.
    10. Developmental – conditions above are present from childhood, continue to manifest themselves in adolescence and adult life.
    11. Delinquency common from an early age.
    12. Aberrant and/or promiscuous sexual behaviour.
    13. Use of substance abuse and alcohol in excess, but use is different from alcoholics and addicts.

17. Alport, G. (1955). *Becoming*. New Haven, CT: Yale University Press.
18. Cleckley, H. (1941). *The Mask of Sanity*. St. Louis: Mosby.

19. Hare, R. D. (1980). A research scale for the assessment of psychopathy in criminal populations. *Personality & Individual Differences* 1 (2), 111–119.

20. Hare, R. D. (1991). *Manual for the Hare Psychopathy Checklist-Revised.* Toronto: Multi-Health Systems.

21. Hare, R. D. (2003). *Manual for the Hare Psychopathy Checklist-Revised* (2nd ed.). Toronto: Multi-Health Systems.

22. It is possible to assess psychopathy in the absence of an interview with the client, provided there is sufficient collateral file information. Wong, S. (1988). Is Hare's Psychopathy Checklist reliable without the interview? *Psychological Reports* 62, 931–934.

    But see: Serin, R. C. (1993). Diagnosis of psychopathology with and without an interview. *Journal of Clinical Psychology* 49 (3), 367–372.

    And note that generally to do the Psychopathy Checklist without an interview requires substantial detail in the collateral files.

    Grann, M., Langstrom, N., Tengstrom, A., & Stalenheim, E. G. (1998). Reliability of file-based retrospective ratings of psychopathy with the PCL-R. *Journal of Personality Assessment* 70 (3), 416–426.

23. The official scoring criteria for the items on the Psychopathy Checklist are protected by copyright and cannot be disclosed in detail in this book.

24. Offenders who commit sex crimes are particularly difficult to score on psychopathy. This is especially true in the civil commitment procedures where sex offenders are being considered for lifelong commitment for future dangerousness. Clinicians have to be properly trained in the assessment of psychopathy if they are going to get the diagnosis correct and make the most accurate recommendation. Note that if a sex offender doesn't score high on psychopathy, that does not mean he is not a high risk to reoffend. Other risk factors need to be considered. And certainly as in the case of sex crimes against children, offenders who meet criteria for paedophilia may pose a different risk for reoffending than those who do not meet criteria for a paraphilia.

25. Hare, R. D. (1998). The Alvor Advanced Study Institute. In D. J. Cooke et al. (Eds.), *Psychopathy: Theory, Research and Implications for Society.* Dordrecht, The Netherlands: Kluwer Academic Publishers. See Livesley, W. J., & Schroeder, M. L. (1991). Dimensions of personality disorder: The *DSM-III-R* Cluster B diagnoses. *Journal of Nervous and Mental Disease* 179 (6), 320; Widiger, T. A., & Corbitt, E. M. (1995). Antisocial personality disorder. In W. J. Livesley (Ed.), *The DSM-IV Personality Disorders.* New York: Guilford Press.

26. Widiger, T. A., Cadoret, R., Hare, R. D., Robins, L., Rutherford, M., Zanarini, M., et al. (1996). *DSM-IV* antisocial personality disorder field trial. *Journal of Abnormal Psychology* 105 (1), 3–16.

Chapter 3: The Assassins

1. From 2000 to 2010 there were over 16,000 murders per year in the United States. Psychopaths constitute approximately 20 percent of the prison population, and using a linear estimate this would indicate over 3,000 murders are committed by psychopaths each year. Note that some studies suggest that over 44 percent of murders of law enforcement personnel in North America are committed by psychopaths, and Hare estimates 50 percent of violent crime is due to psychopaths (http://en.wikipedia.org/wiki/Hare_Psychopathy_Checklist).

2. Hayes, H. G., Hayes, C. J., Dunmire, A. J., & Bailey, E. A. (1882). *A Complete History of the Life and Trial of Charles Julius Guiteau, Assassin of President Garfield*. Philadelphia: Hubbard Bros.

3. Ogilvie, J. S. (1881). *History of the Attempted Assassination of James A. Garfield*. New York: J. S. Ogilvie & Company.

4. Cleckley, H. (1950). *The Mask of Sanity* (2nd ed.). St. Louis: Mosby, pp. 370–371.

5. Johns, J. H., & Quay, H. C. (1962). The effect of social reward on verbal conditioning in psychopathic and neurotic military officers. *Journal of Consulting & Clinical Psychology* 26, 217–220.

6. *New York Times*, July 2, 1882, interview with Mr Harold Emmons, lawyer who had lent office space to Guiteau.

7. Hayes et al., *Complete History*, p. 83.

8. Hayes et al., *Complete History*, pp. 116–117.

Chapter 4: The Psychopath Electrified

1. According to the 2011 US Census, there were approximately 68,873,400 adult males (ages 18–50) in the United States. If 1 in 150 men meets criteria for psychopathy there are 459,156 male psychopaths in the United States. The US Department of Justice reports that there are 2,266,832 inmates in the US jail and prison systems, or about 1,772,073 males ages 18–50 after subtracting females, juveniles, and men over age 50. Assuming 20 percent of these inmates meet criteria for psychopathy, then approximately 354,414 psychopaths are incarcerated or 77 percent of psychopaths in the United States are in prison. Note that men over 50 can be psychopaths, but research shows the majority of antisocial behaviour occurs prior to age 50.

2. Szymanski, M. D., Bain, D. E., Kiehl, K. A., Pennington, S., Wong, S., & Henry, K. R. (1999). Killer whale (Orcinus orca) hearing: Auditory brainstem response and behavioral audiograms. *Journal of the Acoustical Society of America* 106 (2), 1134–1141.

3. Sutton, S., Braren, M., Zubin, J., & John, E. R. (1965). Evoked-potential correlates of stimulus uncertainty. *Science* 150 (700), 1187–1188.

4. Halgren, E., & Marinkovic, K. (1996). General principles for the physiology of cognition as suggested by intracranial ERPs. In C. Ogura, Y. Koga, & M. Shimokochi (Eds.), *Recent Advances in Event-Related Brain Potential Research* (pp. 1072–1084). Amsterdam; New York: Elsevier.

Kiehl, K. A., Stevens, M. C., Laurens, K. R., Pearlson, G., Calhoun, V. D., & Liddle, P. F. (2005). An adaptive reflexive processing model of neurocognitive function: Supporting evidence from a large scale ($n = 100$) fMRI study of an auditory oddball task. *Neuroimage* 25 (3), 899–915.

Chapter 5: The Psychopath Magnetized

1. Yamaguchi, S., & Knight, R. T. (1993). Association cortex contributions to the human P3. In W. Haschke, A. I. Roitbak, & E.-J. Speckmann (Eds.), *Slow Potential Changes in the Brain* (pp. 71–84). Boston: Birkhauser.

2. Good, C. D., Johnsrude, I. S., Ashburner, J., Henson, R. N., Friston, K. J., & Frackowiak, R. S. (2001). A voxel-based morphometric study of ageing in 465 normal adult human brains. *Neuroimage* 14 (1, Pt. 1), 21–36.

3. Ross, E. D. (1981). The aprosodias. Functional-anatomic organization of the affective components of language in the right hemisphere. *Archives of Neurology* 38, 561–569.

4. Winner, E., & Gardner, H. (1977). The comprehension of metaphor in brain-damaged patients. *Brain* 100, 717–729.

Gardner, H., & Denes, G. (1973). Connotative judgments by aphasic patients on a pictorial adaptation of the semantic differential. *Cortex* 9 (2), 183–196.

Cicone, M., Wapner, W., & Gardner, H. (1980). Sensitivity to emotional expressions and situations in organic patients. *Cortex* 16, 145–158.

5. Kwong, K. K., Belliveau, J. W., Chesler, D. A., Goldberg, I. E., Weisskoff, R. M., Poncelet, B. P., et al. (1992). Dynamic magnetic resonance imaging of human brain activity during primary sensory stimulation. *Proceedings of the National Academy of Sciences of the United States of America* 89 (12), 5675–5679.

Chapter 6: Bad Beginnings

1. According to https://www.cia.gov/library/publications/the-world-factbook /geos/xx.html, there are 252 worldwide births per minute (one per 2.4 seconds). A conservative estimate puts the rate of psychopathy in the general population at .05 percent. This means a psychopath is born every

47.6 seconds. Of course, if we have underestimated the rate of psychopathy in the general population (some have argued it is 1 percent), then this birth rate could be much higher. Site last visited 11/11/11.

2. MacDonald, J. M. (1961). *The Murderer and His Victim.* Springfield: C. C. Thomas. See also MacDonald, J. M. (1963). The threat to kill. *American Journal of Psychiatry* 120, 125–130.

3. Bradley, W. E. (1986). Physiology of the urinary bladder. In P. C. Walsh, B. F. Gittes, A. D. Perlmutter, et al. (Eds.), *Campbell's Urology* (pp. 129–185). Philadelphia: W. B. Saunders.

Bradley W. E., & Andersen, J. T. (1977). Techniques for analysis of micturition reflex disturbances in childhood. *Pediatrics* 59, 546.

Bradley, W. E., Rockswold, G. L., Tinim, G. W., et al. (1976). Neurology of micturition. *Journal of Urology* 115, 481.

4. The staff in my laboratory call it the Kiehl-MacDonald Triad hypothesis.

5. Perepletchikova, F., & Kazdin, A. E. (2005). Oppositional defiant disorder and conduct disorder. In K. Cheng & K. M. Myers (Eds.), *Child and Adolescent Psychiatry: The Essentials* (pp. 73–88). Philadelphia: Lippincott, Williams & Wilkins. See also Kazdin, A. E. (2008). Evidence-based treatment and practice: New opportunities to bridge clinical practice research and practice, enhance the knowledge base, and improve patient care. *American Psychologist* 63 (3), 146–159.

6. One could use an auditory version of the Child Psychopathy Scale and simply read the questions to the youth and have them select their responses orally. This would potentially mitigate the illiteracy criticism of self-report inventories.

7. Frick, P. J., & Hare, R. D. (2001). *The Antisocial Process Screening Device.* Toronto: Multi-Health Systems.

8. The Hare Psychopathy Checklist-Youth Version (PCL-YV) is validated only for youth 12 to 18 years old. Researchers are actively examining whether the PCL-YV can be used with younger samples.

9. Fink, B. C., Tant, A., Tremba, K., & Kiehl, K. A. (in press). Assessment of psychopathic traits in a youth forensic sample: A methodological comparison. *Journal of Abnormal Child Psychology.*

10. Frick, P. J., Kimonis, E. R., Dandreaux, D. M., & Farell, J. M. (2003). The 4 year stability of psychopathic traits in non-referred youth. *Behavior Science & the Law* 21 (6), 713–736.

Lynam, D. R., Caspi, A., Moffitt, T. E., Loeber, R., & Stouthamer-Loeber, M. (2007). Longitudinal evidence that psychopathy scores in early adolescence predict adult psychopathy. *Journal of Abnormal Psychology* 116 (1), 155–165.

Munoz, L. C., & Frick, P. J. (2007). The reliability, stability, and predictive utility of the self-report version of the Antisocial Process Screening Device. *Scandinavian Journal of Psychology* 48 (4), 299–312.

Obradovic, J., Pardini, D. A., Long, J. D., & Loeber, R. (2007). Measuring interpersonal callousness in boys from childhood to adolescence: An examination of longitudinal invariance and temporal stability. *Journal of Clinical Child & Adolescent Psychology* 36 (3), 276–292.

11. McMahon, R. J., Witkiewitz, K., & Kotler, J. S. (2010). Predictive validity of callous-unemotional traits measured in early adolescence with respect to multiple antisocial outcomes. *Journal of Abnormal Psychology* 119 (4), 752–763.

12. Burke, J. D., Loeber, R., & Lahey, B. B. (2007). Adolescent conduct disorder and interpersonal callousness as predictors of psychopathy in young adults. [Research Support, N.I.H., Extramural]. *Journal of Clinical Child Adolescent Psychology* 36 (3), 334–346.

13. Lynam, D. R., Caspi, A., Moffitt, T. E., Loeber, R., & Stouthamer-Loeber, M. (2007). Longitudinal evidence that psychopathy scores in early adolescence predict adult psychopathy. *Journal of Abnormal Psychology* 116 (1), 155–165.

14. O'Keefe, T. O., Liddle, P. F., & Kiehl, K. A. (2003, March). *Neural sources underlying emotional lexical decision.* Poster presented at the annual meeting of the Cognitive Neuroscience Society, New York, NY.

15. Levenson, M. R., Kiehl, K. A., & Fitzpatrick, C. M. (1995). Assessing psychopathic attributes in a noninstitutionalized population. *Journal of Personality and Social Psychology* 68, 151–158.

16. Loney, B. R., Frick, P. J., Clements, C. B., Ellis, M. L., & Kerlin, K. (2003). Callous-unemotional traits, impulsivity, and emotional processing in adolescents with antisocial behavior problems. *Journal of Clinical Child and Adolescent Psychology* 32 (1), 66–80.

17. Jones, A. P., Laurens, K. R., Herba, C. M., Barker, G. J., & Viding, E. (2009). Amygdala hypoactivity to fearful faces in boys with conduct problems and callous-unemotional traits. *American Journal of Psychiatry* 166 (1), 95–102.

18. Finger, E. C., Marsh, A. A., Blair, K. S., Reid, M. E., Sims, C., Ng, P., & Blair, J. R. (2011). Disrupted reinforcement signaling in the orbitofrontal cortex and caudate in youths with conduct disorder or oppositional defiant disorder and a high level of psychopathic traits. *American Journal of Psychiatry* 168, 152–162.

Finger, E. C., Marsh, A. A., Mitchell, D. G., Reid, M. E., Sims, C., Budhani, S., & Blair, J. R. (2008). Abnormal ventromedial prefrontal cortex function in children with psychopathic traits during reversal learning. *Archives of General Psychiatry* 65, 586–594.

Chapter 7: Ivy League Lessons

1. According to the 2011 US Census, there were approximately 68,873,400 adult males (ages 18–50) in the United States. If 1 in 150 men meets criteria for psychopathy, there are 459,156 male psychopaths in the United States.

2. Kiehl, K. A., Hare, R. D., McDonald, J. J., & Brink, J. (1999). Semantic and affective processing in psychopaths: An event-related potential study. *Psychophysiology* 36, 765–774.

3. Kiehl, K. A., Liddle, P. F., Smith, A. S., Mendrek, A., Forster, B. B., & Hare, R. D. (1999). Neural pathways involved in the processing of concrete and abstract words. *Human Brain Mapping* 7, 225–233.

4. Kiehl, K. A., Smith, A. M., Mendrek, A., Forster, B. B., Hare, R. D., & Liddle, P. F. (2004). Temporal lobe abnormalities in semantic processing by criminal psychopaths as revealed by functional magnetic resonance imaging. *Psychiatry Research: Neuroimaging* 130, 27–42.

5. Kiehl, K. A. (2006). A cognitive neuroscience perspective on psychopathy: Evidence for paralimbic system dysfunction. *Psychiatry Research* 142 (2–3), 107–128.

6. Harlow, J. (1848). Passage of an iron rod through the head. *Boston Medical Surgical Journal* 34, 389–393.

7. Blumer, D., & Benson, D. F. (1975). Personality changes with frontal lobe lesions. In D. F. Benson & D. Blumer (Eds.), *Psychiatric Aspects of Neurological Disease* (pp. 151–170). New York: Grune & Stratton.

8. Damasio, A. R. (1994). *Descartes' Error: Error, Reason, and the Human Brain*. New York: Grosset/Putnam.

9. Malloy, P., Bihrle, A., Duffy, J., & Cimino, C. (1993). The orbitomedial frontal syndrome. *Archives of Clinical Neuropsychology* 8, 185–201.
   Stuss, D. T., Benson, D. F., & Kaplan, E. F. (1983). The involvement of orbitofrontal cerebrum in cognitive tasks. *Neuropsychologia* 21, 235–248.

10. Schnider, A. (2001). Spontaneous confabulation, reality monitoring, and the limbic system—a review. *Brain Research Reviews* 36 (2–3), 150–160.

11. Anderson, S. W., Bechara, A., Damasio, H., Tranel, D., & Damasio, A. R. (1999). Impairment of social and moral behavior related to early damage in human prefrontal cortex. *Nature Neuroscience* 2 (11), 1032–1037.

Chapter 8: Teenage 'Psychopaths'

1. Source: http://ctjja.org/resources/pdf/factsheet-juvenilecosts.pdf. Annual cost per youth housed in the maximum-security Connecticut Juvenile Training School (CJTC). In addition to the latter annual operating costs, the State of Connecticut spent over $57,000,000 to build the facility, the depreciation of which is not included in the above estimate. It might be relevant to note that Connecticut governor John Rowland served ten months in federal prison for fraud and tax evasion, in part for granting the building contract for CJTC to a construction company in return for free renovations on his vacation home. Site last visited 1/10/12.

2. Across all US states (that reported data), the average annual cost for incarcerating a youth is approximately $100,000 per year in secure residential custody (2011 dollars). This estimate includes youth sentenced to secure group homes, residential treatment facilities, and secure custody facilities. As a youth moves to more secure custody, like maximum-security juvenile prisons, the costs escalate. Source: http://www.justicepolicy.org/images/upload/09_05_REP_CostsofConfinement_JJ_PS.pdf. Average estimate in 2007 was $88,000 per year per youth. Using http://www.usinflationcalculator.com/ this translates to approximately $96,000 in 2011 dollars. Sites last visited 1/10/12.

3. Moffitt, T. E. (1993). Adolescence-limited and life-course-persistent antisocial behavior: A developmental taxonomy. *Psychological Review* 100 (4), 674–701.

4. Forth, A. E., Kosson, D. S., & Hare, R. D. (2003). *The Psychopathy Checklist-Youth Version.* Toronto: Multi-Health Systems.

5. Lynam, D. R., Caspi, A., Moffitt, T. E., Loeber, R., & Stouthamer-Loeber, M. (2007). Longitudinal evidence that psychopathy scores in early adolescence predict adult psychopathy. *Journal of Abnormal Psychology* 116, 155–165.

   Barry, T. D., Barry, C. T., Deming, A. M., & Lochman, J. E. (2008). Stability of psychopathic characteristics in childhood—the influence of social relationships. *Criminal Justice and Behavior* 35 (2), 244–262.

   Frick, P. J., Kimonis, E. R., Dandreaux, D. M., & Farell, J. M. (2003). The 4 year stability of psychopathic traits in non-referred youth. *Behavioral Sciences and the Law* 21 (6), 713–736.

   Obradović, J., Pardini, D. A., Long, J. D., & Loeber, R. (2007). Measuring interpersonal callousness in boys from childhood to adolescence: An examination of longitudinal invariance and temporal stability. *Journal of Clinical Child and Adolescent Psychology* 36 (3), 276–292.

6. Corrado, R. R., Vincent, G. M., Hart, S. D., & Cohen, I. M. (2004). Predic-

tive validity of the *Psychopathy Checklist: Youth Version* for general and violent recidivism. *Behavioral Sciences & the Law* 22 (1), 5–22.

Penney, S. R., & Moretti, M. M. (2007). The relation of psychopathy to concurrent aggression and antisocial behavior in high-risk adolescent girls and boys. [Research Support, Non-U.S. Gov't] *Behavioral Sciences & the Law* 25 (1), 21–41.

Gretton, H. M., Hare, R. D., & Catchpole, R. E. (2004). Psychopathy and offending from adolescence to adulthood: A 10-year follow-up. *Journal of Consulting and Clinical Psychology* 72 (4), 636–645.

Salekin, R. T., Leistico, A. M., Neumann, C. S., DiCicco, T. M., & Duros, R. L. (2004). Psychopathy and comorbidity in a young offender sample: Taking a closer look at psychopathy's potential importance over disruptive behavior disorders. *Journal of Abnormal Psychology* 113 (3), 416–427.

7. Frick, P. J., & White, S. F. (2008). Research review: The importance of callous-unemotional traits for developmental models of aggressive and antisocial behavior. [Review] *Journal of Child Psychology and Psychiatry* 49 (4), 359–375.

8. Viljoen, J. L., MacDougall, E. M., Gagnon, N. C., & Douglas, K. S. (2010). Psychopathy evidence in legal proceedings involving adolescent offenders. *Psychology, Public Policy, and Law* 16, 254–283.

Salekin, R. T. (2008). Psychopathy and recidivism from mid-adolescence to young adulthood: Cumulating legal problems and limiting life opportunities. *Journal of Abnormal Psychology* 117 (2), 386–395.

9. Fazel, S., & Grann, M. (2004). Psychiatric morbidity among homicide offenders: A Swedish population study. *American Journal of Psychiatry* 161, 2129–2131.

Coid, J. (1983). The epidemiology of abnormal homicide and murder followed by suicide. *Psychological Medicine* 13, 855–860.

Wallace, C., Mullen, P., Burgess, P., Palmer, S., Ruschena, D., & Browne, C. (1998). Serious criminal offending and mental disorder: Case linkage study. *British Journal of Psychiatry* 172, 477–484.

10. Nielssen, O., & Large, M. (2010). Rates of homicide during the first episode of psychosis and after treatment: A systematic review and meta-analysis. *Schizophrenia Bulletin* 36, 702–712.

11. Nielssen, O., & Large, Rates of homicide.

Large, M., & Nielssen, O. (2008). Evidence for a relationship between the duration of untreated psychosis and the proportion of psychotic homicides prior to treatment. *Social Psychiatry and Psychiatric Epidemiology* 43, 37–44.

Large, M., Smith, G., Swinson, N., Shaw, J., & Nielssen, O. (2008). Homicide due to mental disorder in England and Wales over 50 years. *British Journal of Psychiatry* 193, 130–133.

Chapter 9: Mobile Imaging

1. See Anderson, D. A. (2011). The cost of crime. *Foundations and Trends in Microeconomics*, 7 (3), 209–265. See also Kiehl, K. A., & Hoffman, M. B. (2011). The criminal psychopath: History, neuroscience and economics. *Jurimetrics: The Journal of Law, Science, and Technology*, Summer 2011 355–397.

2. The not-for-profit Mind Institute in New Mexico changed its name to the Mind Research Network in 2009. I often refer to it as 'The Mind'.

3. Ermer, E., Cope, L. M., Nyalakanti, P. K., Calhoun, V. D., & Kiehl, K. A. (2012). Aberrant paralimbic gray matter in criminal psychopathy. *Journal of Abnormal Psychology* 121 (3), 649–658.

4. Ermer, E., Cope, L. M., Nyalakanti, P. K., Calhoun, V. D., & Kiehl, K. A. (2013). Aberrant paralimbic gray matter in incarcerated male adolescents with 'Callous' Conduct Disorder. *Journal of the American Academy of Child and Adolescent Psychiatry* 52, 94–103.

Chapter 10: The Decompression Chamber

1. Rice, M. E., Harris, G. T., & Cormier, C. A. (1992). An evaluation of a maximum security therapeutic community for psychopaths and other mentally disordered offenders. *Law & Human Behavior* 16 (4), 399–412.

2. Juvenile arrests per 100,000 between 1980 and 2009. http://www.ojjdp.gov/ojstatbb/crime/JAR_Display.asp?ID=qa05201

3. http://en.wikipedia.org/wiki/Stanford_prison_experiment

4. http://en.wikipedia.org/wiki/Abu_Ghraib_torture_and_prisoner_abuse

5. Caldwell, M. F., & Van Rybroek, G. J. (2001). Efficacy of a decompression treatment model in the clinical management of violent juvenile offenders. *International Journal of Offender Therapy and Comparative Criminology* 45 (4), 469–477.

Caldwell, M. F., & Van Rybroek, G. J. (2005). Reducing violence in serious juvenile offenders using intensive treatment. *International Journal of Law and Psychiatry* 28 (6), 622–636.

Caldwell, M. F., McCormick, D. J., Umstead, D., & Van Rybroek, G. J. (2007). Evidence of treatment progress and therapeutic outcomes among adolescents with psychopathic features. *Criminal Justice and Behavior* 34 (5), 573–587.

6. Caldwell, M. F., Vitacco, M., & Van Rybroek, G. J. (2006). Are violent delinquents worth treating?: A cost-benefit analysis. *Journal of Research in Crime and Delinquency* 43 (2), 148–168.

Caldwell, M. F. (2013). Treatment of adolescents with psychopathic features. In K. A. Kiehl & W. Sinnott-Armstrong (Eds.), *Handbook of Psychopathy and Law*. New York: Oxford University Press.

Chapter 11: A Serial Killer Unmasked

1. Du, W., Li, H., Calhoun, V. D., Ma, S., Eichele, T., Kiehl, K. A., Pearlson, G. D., & Adali, T. (2012). High classification accuracy for schizophrenia with rest and task fMRI data. *Frontiers in Human Neuroscience* 6 (145), 1–12.

2. Calhoun, V. D., Maciejewski, P. K., Pearlson, G. D., & Kiehl, K. A. (2008). Temporal lobe and 'default' hemodynamic brain modes discriminate between schizophrenia and bipolar disorder. *Human Brain Mapping* 29, 1265–1275.

3. Ermer, E., Kahn, R. E., Salovey, P., & Kiehl, K. A. (2012). Emotional intelligence in incarcerated men with psychopathic traits. *Journal of Personality & Social Psychology* 103 (1), 194–204.

4. Note that there is no answer that could justify the murders Brian Dugan committed. My quest was in search of his rationale or logic for killing. By understanding his motivations, we are in a better position to potentially remediate these motivations in cognitive behavioural treatment programmes like that of the Mendota Juvenile Treatment Program.

5. Burns, J. M., & Swerdlow, R. H. (2003). Right orbitofrontal tumor with pedophilia symptom and constructional apraxia sign. *Archives of Neurology* 60 (3), 437–440.

6. The US Supreme Court used the term *mentally retarded*, but this terminology has fallen out of favour in academic circles. I have used instead the term *individuals with low IQ*.

7. *Thompson v. Oklahoma*, 487 U.S. 815 (1988).

8. Harenski, C. L., Harenski, K. A., Shane, M. S., & Kiehl, K. A. (2010). Aberrant neural processing of moral violations in criminal psychopaths. *Journal of Abnormal Psychology* 119, 863–874.

9. Callicott, J. H., Ramsey, N. F., Tallent, K., Bertolino, A., Knable, M. B., Coppola, R., et al. (1998). Functional magnetic resonance imaging brain mapping in psychiatry: Methodological issues illustrated in a study of working memory in schizophrenia. *Neuropsychopharmacology* 18 (3), 186–196.

10. Ermer, E., Cope, L. M., Nyalakanti, P. K., Calhoun, V. D., & Kiehl, K. A.

(2012). Aberrant paralimbic gray matter in criminal psychopathy. *Journal of Abnormal Psychology* 121 (3), 649–658.

Ermer, E., Cope, L. M., Nyalakanti, P. K., Calhoun, V. D., & Kiehl, K. A. (2013). Aberrant paralimbic gray matter in incarcerated male adolescents with 'Callous' Conduct Disorder. *Journal of the American Academy of Child and Adolescent Psychiatry* 52, 94–103.

11. *Control subjects* included healthy normal controls recruited from the general population and incarcerated nonpsychopathic control subjects recruited from the prison population.

Index